AT SPEED

UP CLOSE AND PERSONAL WITH THE PEOPLE, PLACES, AND FANS OF

NASCAR

monte dutton

 BRASSEY'S / WASHINGTON, D.C.

Most of this book is derived from material first published in the *Gaston Gazette* of Gastonia, North Carolina. Some of the material was syndicated by Universal Press Syndicate, a division of Andrews McMeel Universal.

LIBRARY OF CONGRESS CATALOGING-IN-PUBLICATION DATA

Dutton, Monte.
  At speed : up close and personal with the people, places, and fans of NASCAR / Monte Dutton.
    p. cm.
  Includes index.
  ISBN 1-57488-261-9 (alk. paper)
  1. Stock car racing—United States.   2. NASCAR (Association)
I. Title.
GV1029.9.S74 D88 2000
796.72'0973—dc21   99–086478

Printed in Canada on acid-free paper that meets the American National Standards Institute Z39-48 Standard.

Design and composition by Melissa Ehn at Wilsted & Taylor Publishing Services

Brassey's
22841 Quicksilver Drive
Dulles, Virginia 20166

First Edition

10 9 8 7 6 5 4 3 2 1

FRONTISPIECE: Dale Earnhardt tilting onto two wheels through the twists and turns at Sears Point Raceway, Sonoma, California, during the Save Mart/Kragen 350, June 1999. *Tom Whitmore*

Dedicated to W. Keith Richardson,
who taught me to be a man,
to do what I think is right,
and not to be afraid of
the consequences of honesty

• • • • • • • • • • • • • • • • contents

Preface **xi**

Introduction **1**

chapter 1 **PEOPLE**

Bobby Allison *Confidence Personified* **5**

John Bickford *The Man Who Gave the World Jeff Gordon* **7**

Geoff Bodine *Learning the Hard Way* **9**

The Burton Brothers *Cut from Separate Molds* **12**

Ricky Craven *King of New England* **13**

Junie Donlavey *The Little Man from Richmond* **15**

Dale Earnhardt *The Fitful Descent* **17**

Dale Earnhardt Jr. *More than Daddy's Image* **19**

Bill Elliott *In Search of Old Glory* **22**

Ray Evernham *The Kid's Protector* **24**

Red Farmer *The Man Who's Seen It All* **26**

Jeff Gordon *Something Completely Different* **29**

Robby Gordon *Forever Young* **30**

Bobby Hamilton *The Two That Got Away* **32**

Ernie Irvan *A Driver Goes to Work Every Day* **34**

Kenny Irwin *The Curse of "The Next Jeff Gordon"* **37**

Dale Jarrett *Never a Wrong Move* **39**

Junior Johnson *The Genuine Article* **41**

Michael Kranefuss *End of a Long, Hard Climb* **42**

Bobby Labonte *Out on the Track, It's Computer Boy* **45**

Terry Labonte *The Iceman Driveth* **46**

Dave Marcis *The Dinosaur at Dusk* **49**

Mark Martin *Nose to the Grindstone* **50**
Jeremy Mayfield *The Southern Alternative* **52**
Lee Morse *He Never Had a Chance* **54**
Jerry Nadeau *With a Little Help from a Friend* **55**
Cotton Owens *Still Plying His Trade* **58**
Adam Petty *The Same Bright Star Still Shines* **60**
Lee Petty *Forgotten Legend* **63**
Richard Petty *The King* **68**
Ricky Rudd *Fighting the Good Fight* **70**
Felix Sabates *Don't Stop Him, He's Rollin'* **73**
Mike Skinner *Lone Wolf* **74**
Jimmy Spencer *He Always Puts on a Show* **75**
Billy Standridge *The Sport Needs Little Guys, Too* **77**
Tony Stewart *On the Charts with a Bullet* **79**
Kenny Wallace *Mr. Nice Guy Feels the Heat* **80**
Rusty Wallace *It Gets to You After a While* **82**
Darrell Waltrip *He Still Loves Racing* **84**
Michael Waltrip *No More Excuses* **86**
Humpy Wheeler *The Visionary as Traditionalist* **88**

chapter 2 PLACES

Atlanta Motor Speedway **91**
Bristol Motor Speedway **93**
California Speedway **94**
Darlington Raceway **96**
Daytona International Speedway **97**
Dover Downs International Speedway **99**
Indianapolis Motor Speedway **101**
Las Vegas Motor Speedway **103**
Martinsville Speedway **104**
Michigan Speedway **106**
New Hampshire International Speedway **108**
North Carolina Speedway **110**
Sears Point Raceway **111**
Talladega Superspeedway **113**
Texas Motor Speedway **116**
Watkins Glen International **117**

chapter 3 **FANS**

They Even Invade the Ballparks **120**
They Love Their Sponsors **122**
Waiting for the Second Coming of Dale **123**
They Are a Hardy Lot **126**

chapter 4 **OPINIONS**

What Do the Drivers Know? **128**
Bruton and NASCAR **130**
The Alternative Is to Believe They're Better **131**
Fall: When the Big Boys Freeze Up **133**
Are the Races Fixed? **135**
Hypocrisy Made This Sport What It Is Today **138**
Message from NASCAR: Two Leagues Won't Work **140**
Hall of Famers? I Think Not **141**
Hey, NASCAR, Let 'em Play **143**
How Old Is Too Old? **145**
The Champ Ought to Be the Biggest Winner **149**
Charlotte's Place Is Secure—For Now **151**
Never Let the Rules Get in the Way of the Show **153**
Road Racers Face Long Odds **155**
NASCAR Takes Over Baseball **157**
Since When Did Jack Roush Take Over NASCAR? **159**
Is There Still Room for Short Tracks? **161**
Rule Book? What Rule Book? **165**
Patches on the Uniforms? Why Not? **167**
Miracles Happen, and Not Always for the Best **169**

chapter 5 **COLOR**

Why Stop with Naming a Race Track? **174**
Vegas Is a Free-Sin Zone **176**
Rules to Cover NASCAR By **177**
Lost Amidst the Dearborn Engineers **180**
Let's Play the Name Game **182**
I'm More than Just a Gear Head, Awright? **184**

chapter 6 **SCENES**

Racing the Old-Fashioned Way **186**

A Press Conference at Alcatraz? Go Figure **189**

This Ain't the NASCAR I Know **191**

Boredom in the Fast Lane **193**

Me and Tony, Down at the Dirt Track **195**

Roughin' It with Young'uns **196**

Head for the Hills, Matilda! Here Comes NASCAR! **198**

Trade Show in Yankee Land **200**

Back Before the Luxury Boxes **202**

One of Many Sad Stories on a Sun-Drenched Day **206**

Helpless in the Press Box **208**

The Last Hooligan Race **210**

Loose Among the Texans **211**

The Winston Is Intentional Mayhem **213**

The Family That Races Together . . . **215**

It's a Long Way from Wilkesboro to Vegas **217**

A Glimpse at the Future **219**

Index **223**

The Author **237**

• • • • • • • • • • • • • • • • • • • **preface**

This project began as a desperate attempt to capitalize on the burgeoning success of NASCAR stock car racing.

As a journalist, I've had a great time on the beat. At the end of 1992, to my satisfaction, I found I was ill equipped to be a businessman, a politician or a public-relations specialist. My main love was writing, so I decided that, for the rest of my life, it was all I would do. I shut down my old business, and with a couple of small exceptions, my life has gotten progressively better since that moment.

I have been a lifelong fan of auto racing. Thanks to opportunities afforded by my late father, Jimmy Dutton (and Ralph Barnes, who ran the meat market at Dutton's Grocery), I had seen Ned Jarrett win a race at Bristol in 1965, when I was seven years old. I had seen Richard Petty win on dirt. I had seen Buddy Baker win the 1970 Southern 500 in a winged Dodge Charger Daytona. More significantly, I had reveled in David Pearson's mythic performances at Darlington Raceway. Perhaps it is because I was a kid, and because childhood memories tend to carry with them the aura of legend, that I never saw a driver who could match Pearson. I still haven't.

When, in 1993, I had the chance to cover NASCAR part-time for the *Spartanburg Herald-Journal*, I jumped at it. To make a living, I also worked as the sports editor of a twice-weekly newspaper, *The Laurens County Advertiser*, near my home in Clinton, South Carolina. To Jim Fair, the sports editor of the *Herald*, and to Jim Brown, the irascible owner of the

Advertiser, I owe a considerable debt. They provided a foundation for everything that has happened since.

Later, I worked for Hal and Selma Hamrick at *FasTrack*, a weekly that specializes in NASCAR coverage. No one has ever done more—and received less credit for his efforts—than Hal Hamrick, a pioneer broadcaster, track owner and promoter, and publisher whose career has spanned the sport's history. Hal and Selma are, in my mind's eye, inseparable from my blood family. The Hamricks and I were, and are, a great match because we are essentially mavericks with an intuitive distrust of the high and mighty. *FasTrack* evolved into the perfect arena for the innate populism that was bred into me by my obstinate father and compassionate mother. It was with Hal's gracious cooperation that I came to cover the Winston Cup Series for another weekly, the New Jersey–based *Area Auto Racing News*. Lenny Sammons and Earl Krause have been every bit as helpful, cooperative and understanding as the Hamricks.

In June 1996, I became the motorsports writer at the *Gaston Gazette* in Gastonia, North Carolina, which was important because it gave me my first chance to cover every race. Thanks to "NASCAR This Week," the *Gazette*'s syndicated page, my work now appears in nearly 600 publications across the country. Later, Universal Press Syndicate gave me the opportunity to become a syndicated columnist. No one has ever had a better editor than Universal's Greg Melvin. At the *Gazette*, Mark Anderson and his predecessor, Michael Smith, have bravely resisted occasional pressure to rein in their excitable and hyperbole-prone "racin' writer."

Since I made that fateful decision to concentrate on sportswriting, every year has been better than the one preceding it. My writings now appear in the *Gazette*, *FasTrack*, and *Area Auto*, on the GoCarolinas.com and Speednet web sites, and in several magazines, including *Racing Milestones* and *Inside NASCAR*.

But I always wanted to write books.

On the Internet, I discovered The Writers Showplace (*www.writersshowplace.com*), which had been designed to help new writers place their work. Gaylene Givens and Francesca Vrattos have offered invaluable encouragement. My Writers Showplace post put me in touch with a literary agent, Jim Cypher, who subsequently took me on as a client.

The nature of *At Speed*, much of which is derived from pieces originally written for the *Gazette*, coupled with the fast-moving world it describes, means that some of the scenes have become slightly dated. For this reason, each chapter is introduced by the date it was originally written. For instance, Dale Earnhardt's "fitful descent" has ended. The greatest driver of his generation won three races last year and ended a four-year decline in the Winston Cup point standings. Ray Evernham has ended his long association with Jeff Gordon and signed on to mastermind DaimlerChrysler's impending entry in 2001 into the sport. Ernie Irvan, unable to recover from a series of devastating injuries, has retired. Dale Jarrett has won a championship. Ricky Rudd has given up "the good fight," shut down his team and gone to work as a driver for Robert Yates. Tony Stewart has, with astonishing quickness, blossomed into the superstar envisioned.

As this is written, the future growth of the sport seems centered around a negotiated television deal that could quadruple the money derived in the past from various networks. Over the next six to eight years, NASCAR and all its far-flung outposts could rake in as much as $3 billion in additional revenues. Will the additional wealth change the sport? Unquestionably. I have a healthy skepticism about whether the changes will be for the better. Monitoring this issue promises to be my next challenge.

Despite the fact that this is a selection of what I consider to have been my best work, some of my favorite racers are not prominently included. Maybe this means I am uncomfortable writing about such figures. Maybe it means that I just glossed over and neglected quality work. Maybe it means a little of both.

It is pertinent, and I do it somewhat sheepishly, to note the absence of profiles of Wally Dallenbach Jr., Elliott Sadler, Johnny Benson, Ken Schrader, Chad Little, two-time Daytona 500 winner Sterling Marlin, Steve Park, Brett Bodine, Derrike Cope, John Andretti, David Green, Joe Nemechek, Ted Musgrave, Robert Pressley, Kevin Lepage, Rick Mast and others. Also absent are profiles of prominent figures like Robert Yates, Mike Helton, Gary Nelson and Rick Hendrick. While William C. France, the president of NASCAR, is not included in the "People" section, mention of his colossal influence is included in several other places.

In particular, I have erred in omitting specific chapters dedicated to Dallenbach, Marlin, Cope, Andretti and Mast, all of whom are not only fine drivers but also interesting characters. Maybe later. The biggest omission, by the way, is David Pearson.

Maybe I'm not good enough—yet—to write about David Pearson.

AT SPEED

HAMPTON, GA.— Frequent readers are well aware of my penchant for criticizing NASCAR when I think something the governing body does is unwise. I hope they realize I do so out of love, not hate. I have been addicted to motorsports since I was five or six years old.

It must be conceded, however, that NASCAR does a lot of things right and that a fair percentage of the slams it receives these days is born of pettiness and jealousy.

I never really considered this until the 1997–98 offseason when I covered the Peach Bowl, Auburn vs. Clemson. During my rare forays into other sports, I personally feel quite a change in atmosphere. On the Winston Cup circuit, I am one of the regulars. When I go elsewhere, I'm "that NASCAR guy." Sometimes I feel like I have a bull's eye on my chest.

Twice while in Atlanta, I went out to dinner with a group of sportswriters, which, of course, is also a common occurrence on the NASCAR beat. It struck me then that a lot of my colleagues had become harshly critical of stock car racing. It was almost as if they felt threatened by it. Its growth raised the specter that perhaps they might be forced to devote more attention to it; accordingly, they seemed determined to tear it down. It reminded me of the way many mainstream journalists treat soccer. These were sportswriters who entered the profession because of their love for football, baseball and basketball.

It's funny. I know many motorsports writers who never followed the sport until they were assigned by fate to cover it. Many of them grew to love it. One advantage I think I've had is that I have followed auto racing since I was a kid, so I have some historical perspective that others lack. Many of the newcomers to motorsports know the present backward and forward, but they aren't in touch with the greatness of the past. Too often they assume that, just because the Winston Cup Series is bigger than ever in terms of popularity, it is better in terms of competition. This isn't necessarily true. To them, Richard Petty, David Pearson, Bobby Allison, Cale Yarborough, Fireball Roberts, Junior Johnson, Ned Jarrett and others are names, maybe even personalities, but not athletes.

I am old enough to have seen Jarrett win a race and to have seen Petty win on dirt. These are cherished memories.

Members of another group see racing only rarely, and only then because its popularity makes their occasional visits unavoidable. A lot of them apparently resent it.

Frequently they nitpick. They have a grand old time making fun of the Southern dialects and fractured grammar exhibited by many drivers and mechanics. They depict the fans as unrefined, hard-drinking rednecks.

Racers are easy targets, but it is also true that their grammar is not much different from that of football and basketball players, most of whom allegedly have educational advantages over the stock car racers. Actually, I think the graduates of the so-called "school of hard knocks" succeed rather well in NASCAR. Considering that many drivers and mechanics have no more than a high-school education, their level of sophistication, in my opinion, is extraordinary.

In a rather odd way, it seems to me, racing is like baseball. You have to know it to appreciate it. Baseball seems dull and slow to those who don't like it. It is high drama to those who do. Racing is never slow, but it does seem like nothing more than "cars going 'round and 'round for four hours" to its detractors.

On the way to Atlanta Motor Speedway Thursday, I stopped at a tiny sporting-goods store in Jackson, Georgia. While I searched for football jerseys for my nephews, the clerk and I started talking sports. When the topic of what I did for a living came up, the clerk frowned.

"That racing," he said, "I just don't get it."

I made my point about a lot of people feeling the same way about baseball, and he was taken aback. At length, though, he conceded that I had a valid point.

Racing does have one major obstacle in its path to further growth and mainstream status in the American sports mosaic. You cannot force someone to love it. That's true of everything, of course, but I don't believe you can make racing popular with as many people as follow football, baseball and basketball. For one thing, a high percentage of urban residents rely on mass transit to get from one place to another. How are these people going to develop the Great American Love Affair with the Automobile?

If a car is, to you, merely a vehicle to take you from one place to another, then you probably are never going to go nuts about NASCAR. I hazard a guess that a small proportion of the people who buy their cars strictly for practical reasons end up driving their Chevettes and Yugos and Kias to the track.

The advantage motorsports has is the intensity of its fans. NASCAR fans more than make up for their smaller numbers with a devotion to the sport that goes beyond mere fandom. It borders on religion, and so do the feelings they have for the drivers.

I don't believe any athlete today has the same relationship with his fans that a Dale Earnhardt does. Not Michael Jordan. Not Mark McGwire. Not John Elway. Earnhardt is a full-fledged, genuine folk hero in the eyes of his fans. This was once common in an America that celebrated men like Johnny Unitas, Joe DiMaggio and "Choo Choo" Justice.

Maybe this, too, will change in NASCAR. But I hope not.

March 1999

Jeff Gordon celebrating his victory at the
Pennsylvania 500, Pocono Raceway, July 1998. *Tom Whitmore*

chapter 1

PEOPLE

BOBBY ALLISON *Confidence Personified*
december 1997

To Bobby Allison, automobile racing is the quintessential American sport. When he climbed behind the wheel of a stock car, he and the machinery were as one. Shifting into gear and accelerating off into a banked turn was no less natural than racing a friend to the end of the block on foot.

"I was always confident," said this giant of the sport. "Competition is a part of life. Everybody feels competition in some degree of something. Competition in a familiar piece of equipment, such as a car, to me makes the most sense."

Allison, the founding father of NASCAR's famous "Alabama Gang," won 40 of his 84 victories during the 1970s, and in perhaps the widest variety of cars in which anyone has ever achieved such success. He won races in Dodge Daytonas prepared by Mario Rossi; in Ford Torinos and Mercury Cyclones fielded by Holman-Moody; in Chevrolet Monte Carlos, Luminas and Malibus from the shops of Junior Johnson; in the Ford Thunderbirds of Bud Moore; and even in the AMC Matadors of Roger Penske. He also won seven times during the decade in cars that he himself owned.

The biggest difference in the competitions of the 1970s and today, Allison said, is the higher mechanical reliability of the present age.

"From day one, even in the late '60s coming up to the '70s, competition was really tough," he noted. "The attrition was way higher. Six or eight guys might start out running pretty good, or it might be four or five. Then a couple would have trouble, and it would narrow down, narrow down, and then, every once in a while, it would get narrowed down to one. What I'm trying to say is, when the green flag fell at the beginning of the race, there were about as many cars that could win as there are now. The competition was really keen, really tough. But by the end of the race, a lot more cars had fallen by the wayside. That's indicative of where we've come in terms of reliability.

"You take the Southern 500 at Darlington, which then as now was probably the toughest race. I led that race one time until really late in the day and I had a failure with eight, 10, 12 laps to go, something like that, and I still finished third. I spent those last laps in the pits while someone else got the win. That was what happened in those days. That was part of what went on, always."

It was during the 1970s that teams had to make what, at the time, was an intimidating transition. The major automobile manufacturers withdrew the full factory backing that had characterized the 1970s. For the first time in quite some time, teams had to find sponsors capable of paying the bills. Allison had one of the first "consumer sponsors"—most of the early sponsors were automotive-related—when Coca-Cola sponsored the Fords, Mercurys and Chevrolets he drove in the early 1970s.

"You had to find out how to run the teams and how to get by without the direct factory support," Allison said. "Cars became better; suspensions became more modern. All kinds of new technologies came in. You just had to do the things necessary to make you better than the other guy on a given Sunday.

"Sponsors were a lot less demanding than they are today. They asked me to do some things outside of racing but not much, nothing like the commitments of today."

The 1970s was a defining era for NASCAR because of the larger-than-life battles between legendary drivers like Richard Petty, David Pearson, Cale Yarborough and Allison, all in the prime of their careers and all determined to win the lion's share of events.

"I think there were a few guys, their personal imprint was a big contributor to the wins that they got," Allison said. "I think it still really takes that personal touch by the driver today. You've still got to have that good car that answers what the driver asks for."

And of Allison's famous rivalry with Petty?

"Richard was different from Cale and David and myself," Allison said. "He had a pretty strong personal commitment, too. He was in a position to have the really good equipment, the really good team, those kinds of things. He applied himself and did a really good job of taking what he had and dominating with it. When Richard was running good, on the days he could win, he made sure he won.

"He was really good at how he handled himself, and some of the rest of us weren't."

JOHN BICKFORD *The Man Who Gave the World Jeff Gordon*
january 1999

CONCORD, N.C.— A decade or so back, John Bickford actually placed a call to Dale Earnhardt for advice in directing the career of Bickford's stepson, Jeff Gordon.

Earnhardt did not reply in person, but even then the man who would win seven Winston Cup titles had heard of Gordon, the teenager who was tearing up the sprint-car tracks of the Midwest.

The message Earnhardt relayed was this: "Tell him to get off the dirt and learn how to save tires."

Earnhardt could scarcely have realized that this advice might prevent him from setting the all-time record for Winston Cup championships. Earnhardt has been stuck at seven ever since Gordon began dominating the stock car scene during the middle years of this decade.

Bickford, now employed at Action Performance Companies Inc., has given up personal supervision of his stepson, who now has business managers and agents hired to anticipate his every move.

"We've got a great father-son relationship, but Jeff's like most kids," said Bickford. "He got a lot smarter than his dad. He makes great decisions and pretty much runs his own life now."

But Bickford scoffed at any notion that, at the age of 27, Gordon has few worlds left to conquer.

"Last year Jeff didn't win all the races, just 13," Bickford noted. "It's not like he doesn't have anything left to achieve."

This father's proudest moment occurred when Gordon called him during his first year in stock cars. While testing at Charlotte Motor Speedway, Gordon had experienced difficulty finding the knack of getting his car through the third turn. The team had brought in another driver to work with Gordon. Gordon's dogged lieutenant, crew chief Ray Evernham, tirelessly pushed the talented phenom until he figured it out.

"You're gonna like Ray Evernham," Gordon gushed. "He never gives up. He's just like you."

Bickford was happy to set the record straight to those who insist Gordon never "paid his dues." In 1987, Bickford had to take several weeks off the sprint-car circuit to raise enough money to continue. Even earlier, in the summer of 1983, Gordon showed signs of burnout. Bickford sent him to waterskiing school just to acquaint him with the types of things "normal" kids experienced. Bickford recalls three or four times when the pressures became too great on the kid. Each time Gordon "came back stimulated and better focused."

Did the kingmaker envision that his prize pupil would ever come this far?

"Hell, no," Bickford said. "How can a parent imagine that? Maybe in our wildest dreams, yes, but who could have known that, at 27, my kid would have his own jet, a motor home, millions of dollars in yearly earnings. I didn't see that coming. I just had a kid with talent, and I wanted him to see it through."

An early-childhood illustration demonstrated the Bickford style.

"I used to tell Jeff that, if you waddle into the kitchen like a duck, if you conduct yourself like a duck, eventually somebody will think you're a duck. If you act like a race-car driver, maybe someday somebody will actually think you're a race-car driver."

And how.

GEOFF BODINE *Learning the Hard Way*
august 1998

HARRISBURG, N.C.— When Geoff Bodine bought his own team, it just seemed like the thing to do.

That was back in 1993, when Alan Kulwicki had just become the first—and only—combination driver–car owner to win the Winston Cup championship. A driver purchasing his own team was as fashionable then as the multi-car team is now.

What not everyone knew was that Alan Kulwicki had been a rarity. On the one hand, he was a phenomenal driver. On the other, he was a calculating businessman. Those two attributes do not exist hand-in-hand with many people. But when Kulwicki was tragically killed in an April 1993 plane crash, Bodine successfully negotiated with Gerald Kulwicki, Alan's father, and bought the team Kulwicki had led to the 1992 championship.

Bodine found almost immediate success, winning three races in 1994 and one more in 1996. But Bodine never found the consistency; his best finish in the point standings was 16th.

Now, after struggling to maintain control of both his personal life and his business affairs, Bodine is a 49-year-old driver looking for a ride.

During the 1997 season, Bodine, beset with so many debts that he had to look elsewhere for help, sold most of the shares in his team to businessmen Jim Mattei and John Porter. Recently Mattei gained control of the entire team. It is widely believed that Mattei is going to name a driver other than Bodine to drive his No. 7 Ford in 1999.

"He (Mattei) will be the sole owner here shortly," Bodine admitted. "We all know this was the late Alan Kulwicki's team. This kind of puts an end to the Kulwicki era and my era of ownership. If not for some minor glitches along the way, that would still be going. I haven't been officially told I wouldn't be driving this car, but that's what the rumor mill says. I'm looking for another ride."

How things have changed since 1993. Of all the drivers who boldly decided they could advance their careers by plunging into ownership—Geoff Bodine, Darrell Waltrip, Ricky Rudd, Brett Bodine, Joe Nemechek, Kyle Petty—only Rudd has managed to hold on without bringing in other investors. Darrell Waltrip sold out to Tim Beverley. Brett solicited help from Andy Evans. Nemechek sold out to Felix Sabates. Kyle integrated his team with his father Richard's ancestral team.

"I don't know what car owners are thinking," said Geoff Bodine. "I thought I would spend every nickel I had to keep a team going, to buy whatever it took to go fast, whatever we needed. I was a racer. Sometimes that gets you in trouble. Some of the new car owners are looking at it as a business. I don't know whether it's good or bad. I'm really confused."

The common view is that big money—Rick Hendrick, Jack Roush, Roger Penske, Felix Sabates—is forcing little money out of business. It is the age of combining resources and bludgeoning the old-timers—Dave Marcis, Bud Moore, Junior Johnson—out of business.

Despite his own experience, Bodine refuses to subscribe to the notion that the individually owned, single-car team is obsolete.

"I don't agree with that," he insisted. "The resources help, but it's the people, the team. I'm a great believer in timing, that if you come along at the right time, great things happen. If you come along at almost the right time, good things may happen. If you come along at the wrong time, you never get the right team, the right sponsor. Jeff Gordon has come along when it's really great.

"It's just getting more difficult, but I'm not going to blame that on multi-car teams. It doesn't always work. It's taken Jack Roush a number of years to make his multi-car effort work. Robert Yates struggles at times. Hendrick Motorsports, where are the other two teams? A lot of people said, if he can do it, I can do it. Since Alan (Kulwicki), it's become more difficult for driver–car owners to survive. Are they going to keep their advantage? It depends on what NASCAR does with testing, aerodynamics and other rules."

The last thing Bodine needs right now is the arrival of "the next Jeff Gordon."

"I don't believe we need another Jeff Gordon," he said. "Then we'd have two guys beating us every week."

Like aging warriors Marcis and Darrell Waltrip, Dale Earnhardt and Morgan Shepherd, Bodine is not ready to give up the chase.

"I've really thought a lot about this in the last month," he said. "I was talking to D.W. during the sewer-main break (that delayed Sunday's race at Charlotte). He was flying. He was wiping himself down with towels. I said, 'What's got into you? You look beat.' He said, 'Yeah, but I'm running. I'm having fun.' Don't think the old guys still can't go out there and have fun and run fast.

"Your priorities in life change. My kids are growing up. I'm divorced. All I have right now is this racing thing. I'm motivated to drive hard, fast, prove I can still do it. Hell, I drove 400 miles at Dover with no power steering. I certainly don't think about retiring just because my son (Barry) is out there driving. I can't wait to drive out there with him."

THE BURTON BROTHERS *Cut from Separate Molds*

july 1998

LOUDON, N.H.— Yes, at times it does seem as if the outcomes of certain Winston Cup races are *too* perfect.

Some have even suggested that Dale Earnhardt's victory in the 50th-anniversary Daytona 500 and Richard Petty's victory No. 200 in front of national television, God and Ronald Reagan were in some manner rigged by NASCAR, the all-powerful governing body.

The Jiffy Lube 300 proves there is nothing to that theory, advanced by a bunch of briefcase-toting, Volkswagen-driving yuppies, all of whom once voted for Walter Mondale.

The last two July Cup events at rustic New Hampshire International Speedway have been won by Jeff Burton, and, as everyone knows, New Hampshire is clearly a Ward Burton kind of place.

Both drivers grew up in southern Virginia, but Ward could live in New England. He is taciturn, conservative and knows that the correct pronunciation for a synonym for automobile is "cah." Ward spends most of his free time crawling around in the woods, wearing camouflage and making weird sounds to mimic various creatures.

When Ward goes to a sporting goods store, he buys a muzzle loader. Jeff comes home with a Duke Blue Devils basketball jersey.

New Hampshire, where you can't even find a Wal-Mart, would suit Ward quite nicely. They'd name one of the lakes after him. Lake Ward (pronounced "wahd") would make for a nice phonetic break from Winnepesaukee, Ossipee, Sunnapee and Winisquam.

What's more, Jeff Burton's dialect is all wrong. Despite the fact that Ward is his older brother, the two sound nothing alike, which is to say that Ward sounds like something out of the woods when he talks and Jeff sounds comfortable with civilization.

Jeff has red hair, which leads one to assume, correctly, that he is mischievous. He has a rich sense of humor and a raucous laugh. Ward is respectful; Jeff is irreverent. Ward errs on the side of caution. He believes that if you don't have something nice to say, say nothing at all. Jeff thinks about something for a few seconds, decides where he stands, says his peace and doesn't worry about what anybody thinks.

But in New Hampshire, where his performance should make him popular, Jeff Burton might as well be an alien. He laughs too freely, has too much fun, plays too many practical jokes.

Jeff Burton is the type of guy who would slosh bottles of champagne in victory lane. Do that in New England and they'd toss you in jail for disturbing the peace. Ward would put his little boy in his lap and buy milk and cookies for everybody.

The state motto of New Hampshire is "Live free or die." Reportedly most choose the former. But it is a difficult choice.

RICKY CRAVEN *King of New England*
july 1998

LOUDON, N.H.— Ernie Irvan grew up in Salinas, California, but when the Winston Cup Series visits nearby Sonoma, other drivers have fans. In fact, Dale Earnhardt probably has more fans than either Irvan or Jeff Gordon, who grew up in Vallejo.

Here in New Hampshire, we have the Ricky Craven Phenomenon. Craven, from Newburgh, Maine, has never won a Winston Cup race, yet he is more popular than Earnhardt, Gordon, Mark Martin, Bill France, Jerry Seinfeld, Carl Yastrzemski and Bob Cousy combined when stock car racing stars visit the Granite State.

For those not familiar with New England and its innate clannishness, Craven's popularity is unbelievable.

Craven, nursing head injuries since early this season, made his return, naturally, in the Jiffy Lube 300. This is a significant story everywhere. But in New England, where Craven is king, it ranks right up there with a papal visit.

And then, he promptly wins the pole. That's like the Pope coming to dinner.

The local papers, the *Concord Monitor* and the *Manchester Union Leader*, battled all week long for Craven supremacy:

MINUTE BY MINUTE: Autograph sessions dominate Ricky's day

PATRIOTIC DRIVER DONS NEW SHOE STRINGS: Red, white and blue laces accompany Craven return

YOU DA MAN: Experts say Ricky could win first Cup race today

I don't know if the attention is overwhelming for Craven, but it sure mystifies and bewilders the writers who cover the rest of the Winston Cup Series. During the pre-race ceremonies, it was quite common for a dignitary to finish his remarks and, off the cuff, add, "Oh, and by the way, go, Ricky!"

In many ways, this is admirable. There is nothing wrong with pulling for the hometown favorite. Furthermore, it's not like the rest of the country dislikes Craven. From the first time I met him, I was impressed with how articulate and thoughtful he was. Craven, in fact, deserves a lot of credit for getting himself out of his car when he knew he was hurt and making sure he was physically ready to drive again. I can't help but wonder how many drivers have suffered post-concussion syndrome and never told anybody about it. It could well be that dozens of drivers have suffered it and had their careers decline as a result. It took plenty of guts for Craven to do the right thing, when he could have convinced himself that it was no big deal and gone into denial.

"I think there's a level of comfort going back home to a place I've had some success and just being home around friends and family," said Craven. "There may be some additional pressure. . . . The pressure comes from within, and it's there every race I've ever driven. Going home is going to be quite a treat.

"I've got to admit it's the only time each year that I get greater applause than Dale Earnhardt and Jeff Gordon. It's pretty significant for me. There's quite a contrast between my introduction at New Hampshire and, say, Martinsville. If there's such a thing as home-court advantage, maybe it's here, by the support I get."

There is, however, an annoying aspect to the Ricky Craven Phenomenon. New Hampshire International Speedway is a pretty raucous place to begin with. There is way too much cheering in the press box. If Craven ever wins a race here, they may just suspend the rules altogether. Sometimes the local worship of Ricky, King of New England, is a bit much.

A year ago, a colleague was trying in vain to write his race lead while a handful of New England scribes discussed Ricky.

"I don't know what the problem is," said one.

"I know whatcha mean," said the other. "I thought today was Ricky's day."

"He just didn't have the cah."

Finally, my friend could take it no more.

"Hey, Ricky Craven's a nice guy and all that," said the sportswriter, "but as a driver, he's a bum!"

The conversation ended. From about five yards away, my first thought was that my fellow sportswriter was about to be lynched. Instead, the heavens opened, then huge black clouds raced across the speedway. Lightning struck the poor scribe, prematurely ending his career.

I'm kidding.

JUNIE DONLAVEY *The Little Man from Richmond*
june 1998

RICHMOND, VA.— Until I met Junie Donlavey, I doubted the existence of leprechauns.

Donlavey is the good-humored little man from Richmond who has been putting stock cars on the track since 1950, when he ran Bob Apperson in the first Southern 500. Donlavey has never had the big bucks behind him, but he's never arrived at the shop to find it padlocked by the bank, either.

Three-time Indianapolis 500 winner Johnny Rutherford drove a Donlavey-owned Ford. So did someone named Yvon DuHamel. Ernie Irvan and Ricky Rudd chauffeured the No. 90, but so did Steve Perry, Gene Felton and Ed Pettyjohn. In 49 years of racing, champions and misfits alike are bound to come straggling in and out of the doors.

The elements have carved Donlavey's cheeks into a weatherbeaten collection of pink tints. When he talks in his Richmond brogue, he always seems to be leaning forward, the better to share a joke or to confide a secret. He has an aura of good humor about him. When you meet Junie, no one has to tell you what a nice guy he is.

The leprechaun is recovering from some hard times. Three months ago, Donlavey underwent open-heart surgery. Now in the middle stages of his rehabilitation program, he showed up at Richmond International Raceway for the first time Friday.

Funny how no one but Junie ever thinks to say nice things about the press.

"I wanted to tell you all that, when I joined the ranks of being a fan and laying around the house and not knowing what was going on, I found out how the fan operates and what everybody looks for when they don't have the access to being in the garage area," said Donlavey. "I want to thank you for the job that you do, because without the print and the radio and the TV, none of the fans would know what is going on. . . . All I can tell you to demonstrate what a beautiful job you have done is, you've kept me informed more than I knew when I was here."

Leo Durocher said that nice guys finish last, and maybe that is true. In 756 tries, exactly one driver, Jody Ridley, has put a Junie Donlavey car in a Winston Cup victory lane, at Dover Downs in 1981. There have been victories in lesser divisions, and three drivers—Bill Dennis, Ridley and Ken Schrader—have been named Rookie of the Year based on turns in Donlavey's cars.

Cold, ruthless behavior may win races, but even the rich and arrogant come to some crossroads where they need a friend and find no one there. That has never happened to Donlavey, and it never will. He will get by with a little help from his friends for as long as he is able.

Donlavey said his surgeon was "the Robert Yates of the medical profession."

"I got so many cards and letters and flowers and fruit baskets," he added, sounding like a character from a Frank Capra film. "I felt like maybe I had a few friends, but I had no idea that as many people would call me and send me cards and letters. It really made me feel good. It made me kind of think that maybe I hadn't gone through life without accomplishing something because they were really good to me."

DALE EARNHARDT *The Fitful Descent*
may 1997

CONCORD, N.C.— Charlotte Motor Speedway is bubbling over with the sentimental notion that Dale Earnhardt is going to win Saturday's Winston all-star race.

Dale Earnhardt. The Intimidator. Seven-time Winston Cup champion. The only three-time winner of the Winston. Considered by some to be the greatest of them all.

So why should anyone consider an Earnhardt victory to be a surprise?

Earnhardt has won 70 Winston Cup races. Only five men have won more. But the most recent one was in March 1996, 37 races ago. On April 29, Earnhardt turned 46 years of age. He has lived to an older age than his father Ralph, a short-track hall of famer who died of a heart attack at age 45.

There is the perception, increasingly widespread, that the great, fierce Earnhardt is descending fitfully into the twilight of his career. His fans desperately want a last hurrah, and this could be it.

This lion of the high banks is not aging gracefully. He and the man who has taken over control of the pride, Jeff Gordon, are business partners off the track, yet they are increasingly quarrelsome on it. Last week, in the Talladega draft, neither would have thrown the other a lifebuoy had he been drowning.

In each of the last two Winstons (the race was known as The Winston Select in 1995–96), Earnhardt has beaten himself. In 1995, he and Darrell Waltrip, both anxious to revisit the heady splendor of youth, rammed their Chevys side-by-side into turn four at an angle where only one could fit. The cool, collected one was Gordon, then all of 23, who went on to win. Last year Gordon and Earnhardt tangled, unwittingly handing the victory over to Michael Waltrip, of all people.

In the Winston 500 a week ago, it was not the Earnhardt of old who settled for second place without mounting a decent challenge to Mark Martin.

"If I had pulled out, I would have gone to the back," lamented Earnhardt.

No. 1, the "back" to which Earnhardt was referring was fifth place. No. 2, John Andretti was willing to take that risk. Gordon was willing to take that risk.

Earnhardt, the all-time winningest driver at Talladega Superspeedway, was content to accept second place, and it's not like he was running for the title. He is a bajillion points (well, okay, 257) behind. If that doesn't seem like a lot, think of it this way. At 10 races into the season, Earnhardt has only 83 percent of the points of leader Terry Labonte. At his present pace, he would be 822 points behind by season's end.

So Dale is going to have to pick up the pace. And a lot of experts think this is where it starts. This speed palace has been notably kind to the Man in Black in the past.

The Winston is annually the site of much intrigue and drama.

It will be interesting to see what Earnhardt brings to the table this time.

DALE EARNHARDT JR. *More than Daddy's Image*
march 1999

DARLINGTON, S.C.— Every waking hour finds Dale Earnhardt Jr., straining to escape the shadow of his old man, yet everywhere but in the cockpit of a NASCAR Chevy, the two are as different as oatmeal and grits.

Some say Dale Jr. is a mirror image of his dad, the seven-time Winston Cup champion, as a young man. But Earnhardt Sr. was never humble. Even in the old days, when he had time to set aside the corporate rat race and mingle with the little people, Earnhardt Sr. had a certain attitude about him. The cockiness might have been tempered by the twinkle in his eyes, but it was always there.

Now the heir apparent arrives in our midst, carrying the hopes and dreams of Earnhardt Nation in his lanky features. "Little E" lacks the swarthiness of his old man. Even as he tries to carry the legacy on his narrow shoulders, he carries himself with a simple grace.

Maybe Dad the Folk Hero said it best: "He doesn't remind me of me. He ain't that mean yet."

Of all the rambunctious youngsters marching double-time in the direction of Winston Cup glory, Earnhardt Jr., at age 24, comes closest to the torrid pace once set by Jeff Gordon. The great man's son has already won rookie of the year and the Busch Grand National championship, all in the same year. Now this innately humble young man is headed to the top with millions of dollars' worth of beacons and trumpets to prepare his path.

The planned date of his Winston Cup debut is May 30. Outside the track, T-shirts are available declaring this "the countdown until E-Day," when Dale Earnhardt Jr. fulfills his destiny and debuts in American racing's most elite club.

"Every time anybody makes a comment about countdown to E-Day, I try to set 'em straight," said the second coming of Dale. "I remind them that it's countdown to my first qualifying attempt. Everybody knows how difficult it is to make these races. I'm really . . . you could sit there and let it pressure you a little bit. They [his fans] are expecting me to get in that race. Nobody's even thought about us not making it.

"You got to kind of get mad about it, and go, 'Man, that ain't the way it is. I might not make it.' . . . If we miss that race, the world's not gonna end."

In the Busch Grand National series, no one has ever been bigger than Dale Earnhardt Jr., and no one will probably be as big again. As the son of the sport's one true legend, Earnhardt is seen by many as its man of destiny. To the Earnhardt faithful, many of whom despise Gordon, Earnhardt Jr.'s task is to wrest control of this sport back from the priggish Gordon.

The expectations are patently unfair. Between Budweiser, the sponsor of Earnhardt Jr.'s upcoming Cup effort, and Action Collectibles, the mammoth merchandiser, corporate America is gearing up to hype Earnhardt Jr.'s arrival with all the bluster of a Hollywood premiere—a premiere that is playing the entire country, from small-town America to Tinseltown.

Is it overkill?

"You may be correct in that assumption," said Little E, as always modesty personified. "It makes a lot of sense to me.

"The only thing that saves me is, I feel like I'm always getting pointed in the right direction. You know, 'you need to be here, you need to be there.' I've got a lot of good people helping me. It prepares me for things like press conferences and things. I really trust those people. They wouldn't put me in a situation where I'd totally bomb out. I'm a race-car driver, I'm not a spokesperson. I'm not an actor. They're waiting for you to say something stupid, and I'm trying to disappoint them."

In the space of two months, Dale Earnhardt Jr. is supposed to move up to Winston Cup, strap himself into a car that has not yet even been put together, and take on Jeff Gordon, Mark Martin, Jeff Burton, Dale Jarrett, his dad and others—and he is supposed to beat them. Why not put him in charge of rebuilding the Charlotte Hornets, too? Maybe at lunchtime the boy could bring lasting peace to Northern Ireland. Ending hunger might have to wait until the offseason.

The fans perceive a great rivalry between Gordon and Earnhardt Jr., especially since it has never materialized between Gordon, still only 27, and the elder Earnhardt, who will soon be 48.

Maybe it will someday, but at present, nothing could be more absurd.

"It's about beaten to death, but I don't think it," said Dale Jr. "I think it is not to his (Gordon's) level yet. You know what I'm saying? What we've done here, what we're doing, the five races (the number of Cup events Earnhardt is scheduled to run this year), isn't quite worthy of him being concerned with us. He's a champion three times over, he's winning races right and left, ain't nobody can stop him. Who are we to be compared to him? I find it, it's kind of uncomfortable because I don't even see me in the same sentence with him.

"It'd be appalling for me to walk around and say, 'I think I'm gonna outrun him at Charlotte,' because, who could say that? It'll be years before I can look over to Tony Eury and say, 'Tony, I think we're good for the pole.' You don't say that kind of thing, even if you are thinking it, you don't say it. You know what it's like? I don't know if this'll sound right, but it's like, one of y'all [the press] just opening up the door to the old man's coach and just walking in there. 'Hey, man, how 'bout an interview?'"

For readers who are perhaps not briefed on the current state of affairs in the NASCAR Winston Cup Series, such a thing is simply not done. The members of the National Motorsports Press Association could sooner gain unfettered access to the Oval Office than to the interior of the elder Earnhardt's motor coach.

"That's what it's like for me. That's how comfortable it is for me to be be compared to him [Gordon]," he said.

What keeps Earnhardt Jr. going is his love of race cars. Outside the car, the world into which he has been inserted is a zoo, a blur and a rat race, all rolled into one. He is McCloud, the New Mexico sheriff dispatched to the tawdry world of New York City.

"I think the rest of the year's gonna be easier," he said, "but we've got some months that are going to be tough. Last month was wall-to-wall, and October's going to be rough, and of course the time around Charlotte is going to be difficult.

"We [he and his race team] have such a close bond, a close relationship, that I still see the enjoyment out of it. I think I enjoy driving a race car more than anybody on the race track. I just enjoy driving a race car. I can't tell you how much fun it is. It'll be a long time before I get too fed up by everything else to quit," he concluded.

Quit? With so much money to be made? Don't be ridiculous.

BILL ELLIOTT *In Search of Old Glory*
february 1997

Four years ago, Terry Labonte was thought to be washed up. At age 36, it was whispered that he didn't have the fire in his belly anymore, that he was content just to ride around and draw a paycheck. Then Rick Hendrick put Labonte back in first-class equipment, and a lot of people had to eat their words.

Wouldn't it be interesting if the same thing happened to Bill Elliott?

It has been 63 Winston Cup races since Elliott last won. Before that victory in the 1994 Mountain Dew Southern 500 at Darlington, Elliott had gone 52 races without winning.

Elliott does not need any more victories to ensure a spot in the various racing halls of fame. He does not need them to ensure financial security. He needs more victories because they are the reason racers race.

Lo and behold, word arrives from Talladega Superspeedway that Elliott's No. 94 was the fastest car down there last week, so fast, in fact, that Elliott canceled a Daytona test session because he didn't want word to get around. According to a couple of the drivers who were with him, Bill's engine-building brother Ernie has discovered a bit of the old magic.

Hendrick knew Labonte could still drive a race car, and the

more discerning observers of racing now know that Elliott can still get the job done. He remains a marvelous driver trapped in quality equipment that is somehow not quite up to the current cutting edge. Furthermore, he is trying to regain the old magic and has been ever since he left Junior Johnson at the end of 1994 and put his career back in his own hands.

Despite all the discouraging statistics, Elliott remains a big star. His fans resolutely get together each autumn and manage to defeat any plan designed to deprive Elliott of NASCAR's Most Popular Driver designation. They have organized committees to get out the vote, whether by paper ballot or 900 phone call. No other driver could weather such a streak of mediocrity without seeing his fan base decline. If Elliott reels off a couple of wins this spring, it will seem like the biggest thing in racing since banked turns.

Elliott has won 40 races. He spent most of the 1980s swapping victories with Dale Earnhardt, with whom he had a rivalry similar to the old Petty-Allison squabbles of the early 1970s. What made Elliott so special? One characteristic that endeared him to fans was his brand loyalty to Ford. Another was the success of his family team in an age when the sport was growing ever more corporate. Then there were the small-town roots, the thick accent, the lanky build and the red hair.

He is never going to be Huck Finn again. Like any successful man, the years have robbed him of his innocence. They have also mellowed him. Never beloved by the press during his prime, Elliott now understands the price of fame.

Quietly he has built a race team that, on paper, ought to win. Only a few teams really go the extra mile with research-and-development programs, test drivers and a whole agenda of plans to provide for the future.

Last year injuries wrecked the Winston Cup effort. At Talladega in April, Elliott's McDonald's Ford skittered through the infield grass, leaped into the air and splattered to the ground with a force that shattered Elliott's left leg at the base of the hip. The crash came on the heels of a string of devastating family tragedies.

After the crash, Elliott came back to race a month ahead of

his doctor's schedule, in part because the necessary rehabilitation had been so harrowing. "I just didn't think I needed to be cooped up at home anymore. If I had stayed home much longer, you could have written me off," he said at the time.

Meanwhile, the performance of young Ron Barfield, Elliott's protégé and heir apparent, provided further evidence of just how shrewd an observer of talent Bill Elliott is. Barfield and Elliott test together, then Elliott sends Barfield out to the Craftsman Truck Series, the Busch Grand National division and the ARCA SuperCar Series to show the world what he has learned.

Eventually, all the data gathered up by Barfield will be reflected in Elliott's Cup performance. Elliott has done his level best to get the program back up to speed. He has quality people, the necessary equipment and enough money to match up with the Childresses, the Hendricks, the Yateses and the Roushes.

In the long run, the luck evens out and the dedicated prevail. When you're filing away your Daytona 500 long-shot picks, don't forget Once-Awesome Bill.

RAY EVERNHAM   *The Kid's Protector*
may 1998

CONCORD, N.C.— Ray Evernham's words cut like a knife.

Evernham and I have quarreled a bit. He was fond of saying I didn't give his driver, Jeff Gordon, enough credit. I responded once by asking how I could possibly give him more credit . . . start a church in his honor? Anyway, in time, I realized that Ray was basically a good guy, a bit overprotective of his driver perhaps, but overall one of stock car racing's more intriguing personalities.

I'd have to say now that Evernham is a bit paranoid, also. For years he has been praised for his intelligence and his role as combination teacher, coach and friend to Gordon.

But Evernham made a mistake last Saturday in the Winston. He let Gordon run out of gas on the last lap of a race in which fuel should not have been an issue. Evernham received some criticism; in response, on Thursday he blasted not only his critics in the press but the press in general.

I can't speak for every story that crossed Evernham's desk, but I didn't see anything myself that was overly harsh. I mean, the team blew the race. It was in the bag when Gordon inexplicably ran out of gas. In 1994, when Gordon won the Coca-Cola 600, Evernham's decision to pit for two tires instead of four made the difference, and Evernham probably received as much credit as Gordon.

Some gotta win. Some gotta lose.

On the one hand, Evernham said the mistake was his fault. On the other, he intimated that the media had no business criticizing him.

"You look at most of the people who gave me a hard time, and they've never done anything in their lives, so it doesn't make any difference to me. . . . It's a lot easier to hit the keys on a typewriter than it is to win a Winston Cup race," Evernham said.

First I was shocked. Then I felt hurt. Then the words hit my insides like ground jalapenos, and the more I thought about it, the madder I got.

During the past year, I devoted some time to establishing a working relationship with Evernham. Obviously that time was wasted. He doesn't seem to have any respect for me or anyone else with a press pass hanging around the neck. Like many in the general public, Evernham thinks we're slime.

Maybe we are. I don't feel like slime, but I'm not objective. Maybe Evernham has compiled a dossier on what each of us "has done," and it reveals nothing. We certainly, at the very least, have a major image problem.

I wish I could introduce Evernham to some of my colleagues. Several served their countries with valor in the Vietnam War. Others were once athletes themselves and played on championship teams. Some worked their way through college. Others' walls are covered with awards that, to them and their families, mean as much as winning the Winston.

I know sportswriters who have performed remarkable acts

of loyalty to friends. Maybe the integrity of journalists is impossible for the Ray Evernhams of the world to understand. His comments didn't really anger me. I felt sad that he has such a low opinion of me and my colleagues. It is hard for me to understand why I can maintain a healthy level of self-esteem, while all about me there are apparently mechanics and drivers who think all of us in the media center are lower than maggots on a piece of raw meat.

Ray used to tell me how much I needed to see the side of Jeff (Gordon) that he saw regularly.

I wish Ray could see my side of the media. I wish he could walk a mile in my shoes. I wish he could experience deadline pressure, and conduct interviews, and assimilate all the information into something marginally coherent.

I wish, somehow, he could see that not everyone can do what I do. I wish he could see that he is one of them.

RED FARMER *The Man Who's Seen It All*
april 1998

One does not necessarily have to win the Daytona 500 to be one of stock car racing's all-time greats. When NASCAR released its own list of the 50 greatest drivers earlier this year, among them were short-track stars like the late Ralph Earnhardt, Jerry Cook and Richie Evans, none of whom ever won a Winston Cup or Grand National race.

Another was Red Farmer, the long-time running mate of the Allisons and charter member of racing's Alabama Gang. Charles Lawrence Farmer is still active on the dirt tracks of that state, and although it is a matter of some controversy, the records show that he will turn 66 years old this fall.

Farmer, whose career is as old as NASCAR's, won three National Sportsman titles (1969–1971) and a Modified title (1956). Yet, he never made much of a stab at the sport's major leagues.

Farmer competed in a total of 36 Cup (and Grand National,

its predecessor) events between 1953 and 1975, never finishing higher than fourth.

"When I was racing in the 1950s and 1960s, factory support was similar to what you have with the million-dollar sponsorships of today," said Farmer. "If you were an independent, you didn't have much of a way to go. I was so competitive, and I wanted to win so much, that I'd rather go win a 100- or a 200-lapper on a short track than run 15th or 20th in Cup just to say I was in Winston Cup. That was the way I felt about it. I didn't have a first-class opportunity, so I'd rather go back to short-track racing."

In the early 1960s, Farmer and the Allison brothers, Bobby and Donnie, were constant companions. Once they slept in the same apartment, with their race cars housed at the same Hueytown, Alabama, service station. They would pull out of Hueytown with their race cars on the back of a truck and drive bumper-to-bumper to some far-off short track where there was money to be won.

"Oh, gosh, here comes that Alabama Gang," was the cry when they pulled into Richmond, North Wilkesboro or Martinsville, and a nickname was born. Neil Bonnett and Davey Allison, both now deceased, were latter-day members of the gang.

"When they had the factories in racing, they pretty well pulled your strings like a monkey," Farmer recalled. "They dominated your life, and that was not what I wanted. When I finished racing somebody, I wanted to sit down and have a beer with him. I didn't want to be told not to talk to him if he was driving the other make of car.

"There were so many great drivers. I wouldn't take anything for the times back in the 1950s when I got to run the (Daytona) beach against (Fireball) Roberts, (Joe) Weatherly, (Buddy) Shuman, Buck Baker, and I wouldn't take any of it back. From the list of the 50 best drivers, I know I ran against 49 of them. I'm not sure if I ever raced against Red Byron or not, but in their era, in what they drove, they were every bit as good as the drivers out there today. I've always said A. J. Foyt might have been the greatest who ever lived because he could drive anything from a wheelbarrow to a semi and make it win. Curtis (Turner) was the greatest dirt-track driver I ever saw, but you can't always judge a driver by his performance, because some of them never get into the best equipment."

Farmer survived the helicopter crash that killed Davey Allison. The tragedies that have befallen that family touched him deeply. His closeness to the Allisons alone, not to mention Bonnett, would be enough to ensure his prominence in the history of the sport.

He has a story to tell about just about everybody. "I remember back in the 1960s, a promoter up in Owensboro, Kentucky, told me he had a local guy blowing everybody away, and he needed somebody to come up there and chop him down to size. We made a deal, and I went up there for a 200-lapper," Farmer remembered. "That guy was Darrell Waltrip. He was a pretty nice guy. He had a big mouth—of course he still has that—but he knew where he was going before he ever left that $\frac{1}{3}$-mile race track."

How did Farmer do against the young upstart? "I ran first, he was second. I took him down a notch, but it wasn't easy."

Now Farmer tows his gold No. 97 to Talladega Short Track, the dirt oval right down the street from the superspeedway, and races every Saturday night.

"I've seen people old at 55 and young at 70," he said. "It's the way you feel, the way you act, what you enjoy doing, that's what determines age. I still enjoy racing, still enjoy it as much as I did 30 years ago."

JEFF GORDON *Something Completely Different*

february 1997

DAYTONA BEACH, FLA.— Each time Jeff Gordon wins a race—and he's won 20 in his last 83 starts—he draws both adulation and scorn from the fans of NASCAR's Winston Cup Series.

Among the ranks of the stock car racing immortals, Gordon is an entirely new commodity.

For instance, Gordon is off for New York City to be a guest on David Letterman's CBS talk show. He likes the irreverent, free-wheeling Letterman style. Dale Earnhardt has always refused Letterman's invitations, preferring instead the more comfortable, guest-friendly Jay Leno.

Gordon, once sequestered by dozens of image makers, is coming into his own, letting his personality show, feeling comfortable with being himself. Now, all of a sudden, it's Earnhardt who seems unwilling to make a public statement without weighing its impact on his souvenir sales.

Like a dozen other Gordon triumphs, the Daytona 500 was the stuff of legend. His bold move on the 189th lap Sunday caused the ever-aggressive Earnhardt to veer outside the bounds of safety, and another daring pass ruined the great Bill Elliott's attempted return to glory. Yes, he had the help of two willing teammates, but his victory in the sport's premier event came at the expense of two of the best drivers ever to turn left.

They grumble about Gordon in the garage area, but even the good old boys are taking notice and, grudgingly perhaps, giving the kid his due.

Dale Jarrett had the best view of the sequence of moves that sent Earnhardt's Chevrolet and Ernie Irvan's Ford rolling down the back stretch. After Earnhardt grazed first the wall and then Gordon—his wheel mark will be seen on the right side of Gordon's No. 24 for the next year as it sits in the Daytona USA exhibit—Jarrett skittered into the rear of Earnhardt, accidentally sending the black No. 3 into its lazy tumble.

"Jeff didn't do anything," said Jarrett. "All he was doing was passing Earnhardt. Earnhardt is the one who lost it."

Was there contact between Gordon and Earnhardt? Jarrett was asked. "Not until Earnhardt hit the wall and came back over and hit us," said Jarrett.

"You can talk about teammates all you want, but it's every man for himself out there. If you get a chance to help each other, that's great. But the object to this game is to win the race."

Not even the great second-generation drivers—Earnhardt, Buddy Baker, Kyle Petty, Jarrett—received a better education in the art of driving race cars than Gordon, a young man raised practically from birth to strap himself into a race car and "let her rip." His stepfather, John Bickford, brought him along step by step, from go-karts to quarter-midgets to midgets to sprint cars to the stock cars that have come to be head of the class in American motorsports.

Now the great child prodigy is growing from robot to fully mature flesh and blood. He has come to understand the responsibility that comes with fame. Already he is astonishingly proficient. Now he can no longer be dismissed as just another cocky young punk with a suitcase full of cash.

ROBBY GORDON *Forever Young*
may 1997

CONCORD, N.C.— Robby Gordon's exciting month of May has not been quite the tour de force that was anticipated.

When these plans were hatched—Gordon is competing today in both the Indianapolis 500 and the Coca-Cola 600—he and car owner Felix Sabates were anticipating a kind of Steven Spielberg production, something along the lines of *Raiders of the Lost Ark.*

Instead, this has turned into a low-budget flick. Call it *Robby's Excellent Adventure.*

Ideally, Gordon, who can swashbuckle with the best of them, was going to win both races. Gordon may be a contender at Indy, where he will pilot something called a G-Force/Aurora. But at Charlotte, the odds against a victory are quite high.

Oh, a Gordon is favored, all right, but it's Jeff, not Robby.

The marriage between Sabates, the multimillionaire car owner, and Gordon, the impetuous young driver, is on the rocks, less than a year after rings were exchanged.

When Sabates hired Gordon, the only sticking point was Indy. Gordon, who had won two CART races in 1995, was unhappy because he found himself trapped in equipment he considered inferior. In addition, the squabble between CART and the Indy Racing League had left Gordon relegated to spectator status for the Indianapolis 500, a race he desperately wanted to win. He decided to defect to NASCAR, but only if Sabates would find a way to run him at Indy.

"I told him (Sabates) that my only regret would be not winning the Indianapolis 500, a dream I'd had since I was a kid," said Gordon, 28. "I told Felix I really wanted to run Indy—I wanted to win Indy—and still try to make a name for myself in a full-time switch over to NASCAR."

Gordon had the reputation as "a wild child" long before he came south, and many doubted he had the patience to wait his turn in the highly competitive world of the NASCAR Winston Cup Series. Sabates has carried the underachiever tag ever since he began fielding a Winston Cup team in 1988.

Gordon's rookie season, to date, has been a disaster. He won a pole early at Atlanta but is yet to finish a race in single digits. Characteristically, he has grown increasingly critical of those around him. After a verbal explosion at Bristol, Sabates took the unusual step of switching the entire teams of Gordon and teammate Joe Nemechek.

Gordon grew increasingly critical of his equipment, and relations between him and Sabates grew strained. When the rest of the Winston Cup Series was competing at Talladega, where Gordon might have been competitive, he elected to pass up the race in order to go for the pole at Indianapolis. He only managed to put his G-Force on the outside of the fourth row.

Dozens of people in the Winston Cup garage would cheerfully lay odds that Gordon and Sabates will not make it to the end of the year together. The only factor that has kept this unholy alliance grudgingly together is the generous terms of the contracts Sabates traditionally extends.

Robby can ill afford to leave, and Felix can ill afford to let him go.

BOBBY HAMILTON *The Two That Got Away*
april 1999

MARTINSVILLE, VA.— At Morgan-McClure Racing, they used to look forward to Talladega.

Not anymore. In Bobby Hamilton, Larry McClure has a driver who has won at Martinsville, at Phoenix and at Rockingham, but never at the track where the No. 4 Chevrolet he is driving has won four times.

In fact, last week Hamilton spoke to an audience of Virginia high-school students and said of the restrictor-plate tracks (Daytona and Talladega), "I don't like them. I think they suck.

"The reason we're on restrictor plates is more of a safety factor for the race fan. Of course, it is a safety factor for the drivers, too. What happens is, these cars are like airplanes. At 192 mph, they're ready to fly. When they get turned around backwards, they actually get in the air. We've got 430 horsepower. Every five horsepower is two mph. If we take the restrictor plate off, it is 735 horsepower. Throw the drag factor in, and we're looking at speeds of 240 mph. Not only would we fly, but we'd land in the top of the grandstands. That's what we've got to be real careful with."

Given his feeling about Talladega, Hamilton, 41, has to be a little glum this week. A lot of pit-road observers thought he had the fastest car going into the April 11 race at Bristol, where he finished 18th. He was 33rd at Martinsville, his favorite track, where he went into Sunday's Goody's 500 as the defending champion.

"I like Martinsville," he said. "Martinsville is a driver's track. You have to be aggressive, but you have to use a lot of race car but not a lot of brake. That's tough to split down the middle. . . . It is a real tough mental situation to run 500 miles here. . . . You've got to take care of the motor and not over-rev it. You've got to take care of the rear-end gear. You've got to take care of the heating problem you have with the radiator. It's a pretty 'conscious' race track. You've got to stay on top of it all the time.

"I like this place, and I love Charlotte. I'm more in favor of the high-banked, high-speed race tracks, but I've always fared well on the flat race tracks. Martinsville is definitely one of my favorites as far as flat tracks go."

Hamilton, who lives in Mt. Juliet, Tennessee, represents the third of four racing generations. His father and grandfather built short-track cars and raced them, and his son, Bobby Jr., is trying to make his break into the truck and Busch series.

"I was in the shop all the time, so racing was in my blood because I was around it. When I got to be 18 years old, I built a car, then I built two cars and I had a driver. Other people don't take care of your equipment like you would, so he tore up a lot of stuff and the very next year I started driving myself. I got to be pretty competitive at the local race tracks, and it had a snowball effect," he explained.

One day Hamilton would like to put Bobby Jr. in the Winston Cup Series.

"I'd have to pick my son, if I had that choice," he said. "If a sponsor wanted a driver who had a lot of experience, where else could you go? You'd get Jeff Gordon if you could. I don't think you could ever do that, but that's what any car owner would want."

At the same time, Hamilton said he understands why the fans get tired of Gordon winning so many races. He feels the same way.

"I don't necessarily hate him," he said. "I get tired of seeing him win all the time. The people booing him are the same people standing at his souvenir rig buying his T-shirts. They get tired of seeing the same person win all the time. That's what's made our sport what it is today, the competitive spirit of the sport, different people winning. We'll see if we can take care of that."

What was a high-school dropout doing speaking to high-school students? Hamilton has made it, but he knows his success defies the statistics. Like almost all high-school dropouts, he regrets it.

"I didn't finish school," he told the kids. "I wish I had. I was fortunate enough to have a dream, and I chased it. I make a lot of money now doing something I love to do, and that is no different than any of y'all can do. If you've got anything in mind out there you want to do in the future, stick to it. It can be done. I'm proof of that. I'm a millionaire now. I'm a redneck millionaire, but I'm a millionaire."

How did Hamilton's school appearance go? "He had the kids eating out of his hands," said Martinsville Speedway public-relations director Steve Sheppard.

What a shame that the race didn't go so well.

"We had a restart and one of the lap-down cars got to the corner and jumped in front of me and locked the brakes," said a weary Hamilton after Sunday's race. "I slid into him and cut down the tire. We lost a lap or two pitting under green, and it hurt the sway bar, too. That pretty much took care of our day right there."

ERNIE IRVAN *A Driver Goes to Work Every Day*
october 1998

DAYTONA BEACH, FLA.— Ernie Irvan, lionhearted as always, climbed into his purple Pontiac and dutifully went out to do battle in Saturday night's Pepsi 400.

Irvan, who had his head rattled on October 11 at Talladega, had seemed incapable of driving just 48 hours earlier. When he drove a handful of laps in Friday night's practice session, it came as a surprise to those who had attended the driver's Thursday press conference.

Ernie, never confused with, say, Peter Jennings in the refined art of public speech, gave a performance that, even by his modest standards, was disquieting. He rambled. He stuttered. His voice seemed slurred, his tongue heavy.

"I'm not in any pain," he had said, "but I've had a headache since Sunday."

What did come through at the Thursday press conference was that Irvan himself did not particularly want to drive in Saturday's race. He talked, as movingly as he could, about his wife and family and about how his relationship with them made racing seem fairly insignificant.

He went out Friday night and completed a few practice laps. When we all got to the speedway on Saturday, there was a release stating that Irvan would at least start the race in his car, with Ricky Craven standing by in case Irvan could not make it.

Some people are going to read this as being critical of Irvan. That is not what is intended. In fact, Irvan deserves credit for his already sternly-established bravery, honed in a battle for life that took more than a year to win after a practice crash in August 1994. Irvan would be the man one would want as a company commander on a battlefield, because he would certainly not shirk a responsibility to bring his men home, whether by dragging them out of harm's way or even by taking a bullet so that someone else could live.

At the same time, stock car racing is not a war to preserve truth, justice and the American way. It's a car race, for gosh sakes, and it doesn't matter that there are millions of dollars involved and millions of fans watching. It's still, in the overall scheme of things, a car race. Certainly, to the men who drive the cars, it is an activity worth risking life and limb for, but not recklessly. Not when a week in bed might make a difference. Not when another bump on the head could affect one's quality of life from now on.

Why was Irvan at the back of the pack, trailing even Darrell Waltrip in the early laps? Because NASCAR's rules say that the driver who starts a race gets the all-important points. Because Irvan was 14th in those point standings and aspired for the prestige that comes with a top-10 finish. Because skipping a race would make that top-10 finish in the points impossible.

Because those NASCAR rules, which make brave heroes strap broken bodies into race cars for no good reason, are idiotic.

Either NASCAR ought to allow a driver to accept substitution when he is injured, or it ought to say that running a few laps at the beginning of a race is insufficient for a driver to earn points.

Drivers have competed with broken legs, broken shoulders and every kind of cracked fibula, tibia and metatarsal imaginable. Tim Richmond competed in Winston Cup races while he had the AIDS virus.

Is it just me, or is some tightening of the rules in order? Might not there be a need for standardized medical requirements?

NASCAR does not have its own medical team, minimum standards for facilities at the tracks or any coherent system of policies and procedures for dealing with driver injuries.

How could they let Irvan race Saturday night? NASCAR officials, when asked, said it was because "he was approved." What constituted approval? Well, he ran a few laps and seemed fine. Doctors let him out of the hospital, didn't they? If he was sick, they'd have kept him. If he'd still been lying in a hospital bed, NASCAR never would have have approved him to drive. That's because NASCAR is in charge, darn it, and what NASCAR says goes.

In view of the incredible growth of the sport, very few things that could happen would put an end to the boom. One of them is if an injured driver went out on the track one starry, starry night and got himself killed. Another is if a driver was allowed back on the track in a car that had already been wrecked and was no longer safe, and he got killed in a wreck.

Irvan came out OK. He drove the first 13 laps, pitted and turned the car over to Craven. Thank goodness. Praise the Lord.

One of these days, NASCAR is not going to be so lucky.

KENNY IRWIN *Curse of "The Next Jeff Gordon"*

september 1998

LOUDON, N.H.— It is easy to look at Jeff Gordon and think his success came effortlessly, that in some way he never "paid his dues."

The truth, of course, is that no one finds success in the Winston Cup Series these days without "paying dues," that Gordon has been paying them since he was zooming around in some go-kart or quarter-midget race at the ridiculous age of five.

It is also true that Gordon never won an official race as a rookie in 1993 and that he won only two in his second season. No rookie has won a Winston Cup race since 1987, when the late Davey Allison won two.

How does one measure competitiveness? In the case of the Busch Grand National division or the Craftsman Truck Series, many vehicles are capable of winning at the beginning of each race, more so even than in Winston Cup. By that definition, the two-feeder series are highly competitive. But if competition is measured by degree of difficulty, Winston Cup is far more competitive than either of the Triple-A leagues. It is extraordinarily difficult to reach the rarefied air where Gordon currently soars. If you don't believe it, ask Ken Schrader or Michael Waltrip.

Or, better yet, ask Kenny Irwin.

Irwin, though slightly older than Gordon, may follow in his footsteps. Irwin's open-wheel accomplishments, driving sprint cars, Silver Crown cars and midgets, are similar in scope to Gordon's. John Bickford, Gordon's stepfather, assisted Irwin in his entry into the stock car ranks. Like Gordon, Irwin arrived in Cup with first-class equipment, succeeding Ernie Irvan (and, indirectly, Davey Allison) in Robert Yates' famous No. 28 Ford.

Irwin is not there yet, and of course, at this point in his career, neither was Gordon.

Thus far in his rookie career, Irwin is 27th in the point standings. In 22 races, Irwin has finished in the top five only once, and he has three top-10 finishes. For much of Sunday's CMT 300 at New Hampshire, Irwin drove like a top-five finisher. He wound up 11th.

"Without a doubt, the level of competition in this series is unbelievable," Irwin said afterward. "The first half of the race, I just knew we were going to finish in the top five, and we didn't even get into the top 10. At the end, my car was the same. Everybody else got better."

In the Truck Series, Irwin picked up a couple of superspeedway victories in 1997. He was seen as such valuable property that Ford Motor Company helped place him with a top-flight ride, the same way Gordon was immediately hooked up with a first-class Chevrolet ride in 1993.

This, however, is not the Truck Series.

"You've got to have so much more stamina," said Irwin. "When I started running stock cars, I was coming off a career in which I was running 30-lap sprint-car features. Up here, these guys run every bit as hard for 300 laps, or 500 miles, or whatever, than I did for 30 laps in [the United States Auto Club]. You gotta be *on it* all the time up here.

"You've got to develop mental toughness, that's what it takes, and you can work out every day for physical toughness, but it's harder to build up mental toughness. I'm learning that, too."

Irwin is making progress, though some would insist that his performance this year is a disappointment. Those are the observers with poor memories. Gordon, the "breakthrough" rookie of this decade, was 14th in the standings in his rookie year, no better than eighth his sophomore season.

Will Irwin improve from now until the end of the season? Almost certainly. Will he win, say, next year? Who can say? Will he be "the next Jeff Gordon"? It's like declaring unequivocably that Dale Earnhardt Jr. will be better than his father.

A Gordon or an Earnhardt does not come along every year, probably not even every decade.

DALE JARRETT *Never a Wrong Move*
july 1997

High above the coastal region of South Carolina known as "the Pee Dee," Dale Jarrett stared out the window to watch as the NASCAR-owned plane in which he was riding raced through the swirling clouds.

Partially because he has a father and a brother who are card-carrying TV personalities, Jarrett is extremely media-friendly. Partially for the same reason, he knows the media too well. Jarrett is a master at media relations: cordial, warm, friendly, quotable. He is also a master at never saying too much. He guards the morale of his race team with the smooth affability of a public-relations representative. The mood of Jarrett can sometimes be read as much by what is not said as by what is.

On this morning, Jarrett flew to Myrtle Beach for a promotional appearance at the NASCAR Cafe. The entire round trip from his home in Hickory, North Carolina, takes no more than five hours. Jarrett arrived at the airport in his maroon Ford Expedition, casually dressed as befits the easygoing Winston Cup celebrity.

Once in Myrtle Beach, Jarrett climbed out of the plane to greet two carloads of representatives of the restaurant, all anxious to share chitchat with the driver of Robert Yates' rapidly-growing-famous No. 88 Ford Thunderbird. Jarrett's overwhelming success has brought back the prestige of the number, once a power when Darrell Waltrip made it stock car racing's first "notably green" car during the 1970s. Yates was a mechanic at DiGard during those days, so when he began a second team in 1996, he stuck an 88 on the side of his new red, white and blue Ford. No. 28, Yates' more familiar number, has been sharing the spotlight ever since.

Unlike most drivers, Jarrett never had to learn the art of being multi-dimensional. Ned Jarrett wanted both of his sons to know about more than turning steering wheels and wrenches. Like his father, Dale is the graduate of a Dale Carnegie course. As a boy, he took up both golf and baseball with notable success. He can talk stocks with a broker or earned run average with a baseball fan, and he's not repeating what someone else told him.

The Stock Car Cafe's ownership/management team is made up of a veritable mafia of University of Tennessee graduates and fans. Upon learning this, Jarrett quickly segued into a discussion of why quarterback Peyton Manning decided to return to Knoxville for his senior season. He pointed out several facts of which the sportswriter in the back seat was unaware.

Later on, Jarrett shared a principle he got from Ned, his father.

"Most people treat you the way you treat them," he said matter-of-factly. "Dad pointed that out to me, and at the time, we were talking about the media, but it's not just with the media. It's with people in general. If you don't treat them with respect, you can't expect to be treated with respect in return."

The perspective of watching an athlete being swamped by fans is not the same as that of fans swamping that athlete. They all wanted a piece of him, shoving forth caps and T-shirts to be autographed, posing with the star three different ways while the black-sheep cousin without enough clout to be in the photograph tried to figure out how to operate the camera, and insisting on conversations despite the fact that similarly persistent fans are lined up 15 deep behind them.

They're nice people; they just went slightly nuts in the presence of their hero.

Meanwhile, their hero had to resist the inclination to go slightly nuts himself. Dale Jarrett did so effortlessly. He is a professional. He agreed to make this appearance. An early-morning round of golf was already out of the question. This was the task at hand, so he decided to enjoy himself.

He did. Not a word of complaint. He asked only for a glass of tea and cheerfully retired to a nearby bathroom to switch his patterned golf shirt for one with a NASCAR Cafe logo on it. He endured every demand on his time gracefully.

Millions of dollars pass in and out of this man's bank accounts. While elevating his lifestyle and ensuring the future security of his family, the money brings with it commitments that endlessly complicate every aspect of his life.

Dale Jarrett works hard at being a hero. It just doesn't look like it.

JUNIOR JOHNSON *The Genuine Article*
october 1997

In 1973, Junior Johnson attended the opening of a movie that was based on his life. *The Last American Hero* starred Jeff Bridges and Valerie Perrine. The movie had been inspired by a magazine article written by *Esquire's* Tom Wolfe.

When someone asked Johnson what he thought of the movie, all he wanted to talk about was his father.

"I enjoyed it more this time than I did the first time I seen it," said Johnson. "I tell you, if you'd knowed my daddy, that couldn't have been no more like him. Now he had his ways, but that was just the way he was."

Robert Glenn Johnson Jr. is a mixture of fact and fiction, which is to say he is not only the last American hero, but one of the last icons of American folklore. Even in the early days, the moonshine origins of stock car racing were a bit overstated, but Johnson was the genuine article, a moonshine runner who had survived to use his talents in a more lucrative, less dangerous, more civically responsible fashion. Johnson spent almost a year in a federal penitentiary in Chillicothe, Ohio. His father "pulled four or five years."

"I got caught at the still," he said proudly. "I never got caught on the road. They never outrun me."

Even during his moonshine days, Johnson never considered himself a criminal. It was a way to make a living in an age and a place where that was a struggle. Then as now, Johnson lived in Wilkes County, North Carolina.

"We thought it was just a law made to take away your right to do what you wanted to do and to make a living," he recalled. "We thought we were as much justified doing what we were doing as the guy who was selling the sugar, the containers, stuff like that. In our section, that was just a way of life. It was just a business."

He is not a criminal. He was pardoned by the President of the United States. He was, first, a fabulous talent, a natural behind the wheel. Second, he was the strong, silent type, never one to volunteer his opinion, a man who communicated by example.

As a driver, Johnson was the greatest charger of all time, a winner of 50 races only beaten in most cases by the limits of his equipment and rivaled in that glorious category only by Curtis Turner, Cale Yarborough, Buddy Baker and Tim Richmond. He retired in his prime and went on to become one of the greater car owners, his victory total exceeded only by the three generations of Petty Enterprises.

Johnson possesses the famous Southern knack of covering up a considerable native wisdom with a penchant for simplicity that others mistake for lack of intelligence. He has been foolishly underestimated hundreds of times.

Near the end of his car-owner career, a Johnson car driven by Jimmy Spencer won the summer race at Daytona. Enjoying the victory in the press box, Johnson let down his guard for one of the few times in his five-decade racing career.

"I remember when people used to say I messed up by hiring Terry Labonte," he said, the syllable dropping off his tongue like sorghum molasses. "I probably made more money off o' Terry Labonte than any driver that ever drove for me.

"I reckon I ain't quite as dumb as folks think I am."

MICHAEL KRANEFUSS *End of a Long, Hard Climb*
june 1998

LONG POND, PA.— It was all a matter of time. That's easy to say now.

Michael Kranefuss, once the mastermind of Ford's entire motorsports program, was destined for success from the moment he retired from the Dearborn, Michigan, corporate offices and started his own Winston Cup team. Like Roger Penske, who eventually became his partner, Kranefuss experienced his share of bumps and bruises after he set up shop in 1994. Jeremy Mayfield won Sunday's Pocono 500 in Kranefuss' 99th race as a car owner.

"We took a few detours," said Kranefuss, "and it took a long time to get the right people together, and that's what really makes the difference. Everything else you can buy.

"If you don't have the driver who is cheering everybody on, keeping everybody in line, if you don't have the crew chief who keeps everybody and everything in control, and if you don't have an owner who is supportive, and I don't mean this in a bragging way, but I mean it takes a few people, and I think this: We have been able to take care of everybody . . . and it showed."

Kranefuss combines the determination and precision that one identifies with his German upbringing with a dry sense of humor that keeps him in good stead with the Southerners and Midwesterners responsible for putting his Taurus on the track week after week.

"I knew we had a great car, but when you think about the way we build the cars, paint them, the way we keep the place clean. Being number one is an attitude," said Kranefuss in the Pocono International Raceway press room.

"After Jeremy took the checkered flag, for a while I was totally numb when I was coming down the stairs (through the grandstands). There was absolutely no emotion."

"Michael Kranefuss, he deserves this," said third-place finisher Dale Jarrett. "He's done a lot for this sport, and I just want to congratulate him."

Kranefuss was asked if it had taken him longer than expected to win. His answer reflected typical candor.

"Sure," he said, no longer numb, "the second race was already too long. I don't think anybody has got an appreciation of how difficult it is to be consistently good and run up front in Winston Cup racing. To get there, you've got to have the people, and you've got to understand NASCAR.

"I've been in every other form of racing, be it Formula One or Indy cars, drag racing, sports cars, rallies. Usually the people who have got a good car on any given day, even if they've got a little bit of a problem, they can still finish well in the points. Here, say you've got a flat tire, you're lucky to finish in the top 20 and sometimes in the top 30.

"At many Winston Cup races that I attended when I was working for Ford, I learned that deep-down understanding you only develop when you are there every day and when you pay for it with your emotions and part of your checkbook, too. It took me a lot longer than I thought it would, that's true."

But when Kranefuss' team arrived, it did not just arrive in victory lane. Mayfield leads the point standings, and he has been at or near the top of the standings long enough to indicate that he is a genuine contender for the championship.

Once Kranefuss' partner was Carl Haas; now it is Penske. Once he thought John Andretti would take his team to the top. When he became convinced that the chemistry wasn't there with Andretti, Kranefuss engineered a bold trade with Cale Yarborough that brought Mayfield on board. Everyone knew that Mayfield had talent. No one, save for Kranefuss perhaps, imagined that Mayfield could come so far so soon. It took a year to put the elusive pieces in place, a year in which Andretti was winning a race at Daytona for Yarborough.

Kranefuss stood by his judgment. On Sunday, he could crow. He even attempted a ridiculous, German-tinged impression of Gordon, who had to settle for second behind Mayfield.

"At the [last] restart, I was listening to the radio, and Jeff Gordon said, 'Man, Ray (Evernham), you better have the 12 car (Mayfield) put on the chassis dyno.'"

Gordon, who on most afternoons is the one drawing the suspicions, was beaten, and he knew it.

"He (Gordon) was talking about the chassis dyno," said Mayfield, "and he stopped me during the (red-flag) break and was talking about motor this and motor that. He didn't give our whole race team credit. I mean, we had an awesome motor [Sunday] and we had an awesome race car, and we showed that through the corners and down the straightaways. Hey, they need to put [Gordon] on the chassis dyno. His car ran just as fast as mine did."

This may have been the first of many struggles between Mayfield and Gordon.

BOBBY LABONTE *Out on the Track, It's Computer Boy*
december 1999

One thing you need to know about Bobby Labonte: He's quirky.

It is said that Jeff Gordon is a phenomenal player of video games. I can't say. But, without knowing anything else, just from observing Labonte and Gordon, I'd put my money on Labonte. He's a little guy (so is Gordon, and so are about three-quarters of the one-time go-kart racers now populating the Winston Cup Series); his eyes dart around like he's looking for monsters or Nazis to fire away at, and when you talk to him, he fires those words out at you.

It's hard to make heads or tails of Labonte in an interview. First, he usually doesn't answer your question. It's not that he's ducking it; he just thinks of a better one for you.

Traditionally, here is the drill. Labonte listens to the question, gets this goofy expression on his face, turns his head, turns it back, rests his chin on the knuckles of his left hand, thinks about it a bit more, and finally says something like:

"Well, I'm not sure I understand the question. What I think you might be saying is, did I do this? Well, I'm not sure whether I did this or I did that, but one thing I *did* do was this other thing. . . . "

Then, off he goes, in some wayward direction. On occasion, he will actually end up answering the original question, but that's more of a coincidence that anything else.

Bobby also runs the option on occasion, or maybe he reverses his field.

"I did this, or maybe I did that. Who knows?"

"Well, he asked me this, but I told him, well, that. But then the more I thought about it, maybe it *was* this. Does that make sense, or does it matter?"

In terms of the next day's story, it doesn't matter.

God help us when he one day discovers the price of eggs in China. It will be all over then.

Labonte is a remarkable driver. No one disputes that, with the possible exception of him, because in conversation, he disputes virtually everything. And he agrees with virtually everything. The odds are about even that he will answer a given question either way. Or both.

Does that make sense? And did I mention that he makes quite an impression on others, to the point that they actually start writing in the same sing-song manner that Labonte speaks? It's a new style of journalism: Labontespeak. Something like that. Maybe not. Then again, maybe.

Later that day, when Labonte's interviewers leave the track, they invariably go by a nearby Wal-Mart and pick up a copy of one of those NASCAR simulation computer games.

That's so they can understand exactly what it is that makes this 35-year-old prodigy tick. But nothing works. He's incredibly intense. We have no idea why. It's like relating to Robby the Robot. Or an extraterrestrial. But that's impossible. He bears some similarity to his older brother. Could he be a droid? Or better yet, could both Labonte brothers be, you know, invaded by body snatchers?

Nah.

But that *would* explain a lot.

TERRY LABONTE *The Iceman Driveth*
december 1999

Terry Labonte is something of a throwback to an earlier era.

After what is perhaps the most famous race in NASCAR history, the 1976 Daytona 500, principal figures David Pearson and Richard Petty were remarkably sportsmanlike. As the two drivers—Pearson in the famous red-and-white Wood

Brothers Mercury and Petty in his even-more-famous No. 43 Dodge—exited the final turn on the final lap, they crashed. Petty's Charger slipped high on the track before he had cleared Pearson's Cyclone, and the two cars gyrated wildly into the infield grass. As most longtime fans know, Pearson managed to keep his car running; Petty's was shut down and stranded, no more than 100 feet from the finish line. Pearson drove his smoldering car across the line, and Petty finished second.

Petty calmly took the blame for the incident. Pearson, from victory lane, said simply, "I don't feel like he done it on purpose." It was just another day at the office for both of these legendary racers.

That's the way Labonte is.

In an age of hair-trigger tempers and relentless name-calling, Labonte clings to the old-time notion that it is better not to air one's dirty laundry in public. It is wrong to conclude, as many have, that Labonte is bloodless. He has the competitive spirit of all athletes, but if he can control his emotions, he will. He will almost always wait until he has cooled down, then emerge from hiding to discuss a situation, no matter how inflammatory, with something approaching cool reason.

Hence his nickname: The Iceman.

Terry Labonte is, to borrow a cliché, a driver's driver. Owing partly to his famous terseness, he has never captured the imagination of great throngs of fans, at least not in the manner of the roguish Dale Earnhardt. Neither has Labonte enjoyed the gaudy success of a Jeff Gordon. He merely does his job, probably as efficiently as anyone ever has. The performance mirrors the personality.

Where Labonte does have a considerable body of support is among the ranks of fellow drivers. He is often cited as a guiding influence by up-and-coming youngsters. One of them, oddly enough, is his brother Bobby, who is seven and a half years younger.

Most of the time, Terry is exceedingly pleasant. He wears a thoughtful smile for most of his waking hours (and probably, for that matter, while he is asleep). He is not much for sound bites, which makes interviewing him a chore, especially for the electronic-media operatives who are accustomed to placing a microphone in front of someone's face and watching the subject ramble.

Labonte does not ramble. To get a good answer, one must produce a good question. The old saying does not fit precisely: Labonte does not suffer a fool gladly, but his suffering is nice and polite. He is devastatingly adept at hinting pleasantly that the question he has just been asked is unworthy of response.

Once, at Rockingham, Labonte entered the media center for a press conference after winning the pole. One of the questioners pointed out that, during Labonte's missing years (he spent four seasons languishing in subpar equipment), he had managed to win a pole only once. Did his recent qualifying resurgence have anything to do with his move to Hendrick Motorsports?

"Well, I haven't been in this sport but a little while," said Labonte, then a veteran of nearly 500 races, "but one thing I've already started to notice is that . . . it really helps to have good equipment."

Labonte betrayed not even a hint of sarcasm in his bland smile, but throughout the room, seasoned reporters were elbowing one another in the ribs and breaking out in laughter. If Earnhardt had been asked the same question, he would likely have growled, cussed and maybe even cut the interview short. It's one of the reasons why Earnhardt doesn't often get asked such questions. Labonte just smiled.

Another Labonte trademark is the smiling no-comment. A reporter will nervously ramble on and on, finally pause long enough for Labonte to flail at some answer, and Labonte will decline. Nicely.

He will shrug his shoulders, shake his head, smile patiently and say, "Uh, I, uh, don't know. Beats me."

It is great theater, watching this latter-day Gary Cooper performing, adding layer upon layer of subtlety, in our midst.

The temper does flare, by the way, but almost always outside of the glare of public view. Once, after an IROC (International Race of Champions) event, he felt as if officials had unfairly allowed Earnhardt to change tires. Rumor has it that Labonte ripped off the uniform provided all participants in that series, wadded it up and stuffed it into a trash can, vowing he would never again take part in such a farce.

By Monday, he had reconsidered, saying he was just a bit frustrated. "No big deal." To hear Labonte tell it, nothing ever is.

DAVE MARCIS *The Dinosaur at Dusk*
january 1998

SKYLAND, N.C.— At age 55, Dave Marcis is a glorious anachronism, as out of place in the modern NASCAR landscape as a moonshine runner or a shade-tree mechanic.

On the wall in Marcis' office is a picture of John Wayne ("He was a guy who told it like it was and who always said what he thought."). Two rather noteworthy slogans are posted on the walls:

"We have done so much with so little for so long that now we can do anything with nothing!"

"The only thing more overrated than natural childbirth is owning your own business."

Marcis Auto Racing has seven employees. Robert Yates Racing has 71. Yates fields two cars, yes, but it hardly takes a mathematician to conclude that Marcis is playing high-stakes poker with a deck loaded against him.

He started 20 of the 32 races last year, probably squandering an average of $12,000 each of the 12 times he failed to qualify. Yet he soldiers on: undermanned, underfunded and underappreciated.

"When I first came to NASCAR, you could be a handyman with a good head on your shoulders, put a car together and do decent with it," Marcis said. "The cars used to not be so sensitive. You could muscle them and be successful just by using your wits."

Marcis landed in the lower slopes of the Blue Ridge in the late 1960s, when he migrated from Wisconsin to make his fortune in the stock car homeland. He fell in with a couple of other plain-spoken bushwhackers, Cecil Gordon and Bill Seifert, and parked his mobile home near theirs. Gordon, from Horseshoe, and Seifert, from Skyland, have long since surrendered their tenuous places among the touring regulars of Winston Cup. Only the obstinate Marcis remains, more a burr under the saddle than a revered elder statesman of the sport.

"I can't say they don't appreciate me anymore," insisted Marcis. "I guess it's just that there's not as much need anymore for guys like me. NASCAR has been good to me. I've had my differences, but I've never been a burden and I've never tried to hurt the sport."

Marcis' sponsor, Realtree, probably chips in less than $2 million, which is about what the big teams get for associate sponsorships. It beats the heck out of nothing, though, and Marcis knows what getting by on nothing is like. He earns outside income by helping Richard Childress Racing on test sessions and by setting up the International Race of Champions Pontiacs.

In the entire history of NASCAR, only Richard Petty has started more races than Marcis. It has been 15 seasons since the last of his five victories. A 15th place at Texas was the highlight of 1997, yet Marcis is no stroker. With seven employees and a motley crew of race-day helpers, it's tough to keep up with the moneyed elite, especially on the way out of the pits.

"I don't just ride around," said Marcis. "I go out there to race. I probably could finish better a lot of the time if I didn't race as hard."

It's not his nature.

MARK MARTIN *Nose to the Grindstone*
may 1997

CONCORD, N.C.— Why is it so often the case that the most spectacular of athletes is burdened with the blandest of personalities?

Perhaps it is because the talented athlete takes his extraordinary gifts for granted.

Mark Martin, suddenly the talk of the Winston Cup Series, is himself not much of a talker.

This short bundle of Arkansas work ethic is, by all accounts, a devout Christian and loyal family man. The sight of Martin spending time with his 4-year-old son, Matthew Clyde, is a familiar one in the luxury-bus enclaves where NASCAR's nouveau riche frequently reside.

Oddly enough, Martin is easier to interview during the hard times than after victories. After one of his many Winston Cup and Busch Grand National triumphs, he almost freezes up, so anxious at times is he to appear humble. Getting him to take credit or rejoice can be next to impossible.

But when Martin is down and out, as during the recently ended 42-race Cup losing streak, he can be thoughtful and provocative. When Martin lets down his guard to talk about his boyhood and his relationship with his father, the words carry a certain poignancy.

Little Mark Martin, always undersized, learned his race-car driving the hard way, in his daddy's lap. Father and son used to burn rubber on the hilly blacktops around Batesville, Arkansas. As Mark readily admits, the law doesn't allow fathers and sons to have fun like that anymore.

"My dad was born too soon," said Martin. "If he'd have ever had a chance to do this, he would have made a lot of these drivers look bad."

Martin puts more pressure on himself than any of his opponents can muster. He professes no satisfaction in running 10th.

"Winning is the only thing that matters to me," he said recently. "It's the only way I know to have fun."

Fun. An interesting term, coming from Martin, who smiles often but seldom laughs. He is gracious to his fans, but it is the workmanlike courtesy of a salesman, not the spontaneous affection of a movie star. Martin is famously punctual and efficient. He manages to maintain a successful career in three major series (Winston Cup, Busch Grand National and IROC), uphold his obligations to his sponsors and fans, and even work out with weights an average of two hours every day.

But there is no apparent passion, only nose-to-the-grind-stone determination. He has the will power of a runt the big kids used to pick on, the staying power of a one-time phenom who fell flat during his first shot at the big time, and the simple faith of a reformed sinner.

Up close and personal, Martin may be the sickliest looking health nut in America. The skin tone is ashen and leathery, completely colorless save for the mild coat of pink from the sun. The rapidly graying, spiky hair makes him look older than his 38 years. He would easily pass for 50. His smile appears to be the creation of a man with a penchant for staring at a mirror and posing for cameras.

But he is a driver's driver, just as capable of winning on the high banks of Talladega as on the twists and turns of Watkins Glen. If Martin says he will do something, his word is as good as gold.

There is much to admire in Mark Martin, but there is little to love.

JEREMY MAYFIELD *The Southern Alternative*
june 1998

LONG POND, PA.— More than one person has written that Jeff Gordon needs a rival from his own age group.

Congratulations, Jeremy Mayfield. Your mission, should you choose to accept it, is to battle the two-time Winston Cup champion week after week, Ford against Chevy, tooth and nail.

After winning his first Winston Cup race, Mayfield indicated he was willing to take up the mantle.

"As we would say in Kentucky, 'Bring it on,' " said Mayfield, thereby breathing a ton of fresh air into the Pocono press room.

If Mayfield plays his cards right, he will not have to duck

beer cans and face the weekly boos. No one is going to sell F.A.G. (Fans Against Gordon) T-shirts about Mayfield, who is just a couple of years older than Gordon and, in more than one way, everything Gordon is not.

If Gordon's detractors knew him, they would be ashamed at the way they boo him, spread rumors about his personal life and otherwise leave the country's most burgeoning sport dulled by an unsavory pallor.

Yet one of the reasons they hate Gordon so is that he comes across as plastic and programmed. Too often Gordon says what his handlers have told him to say. The good old boys also want more from Gordon than he is willing, or able, to give. They want him to do more than just drive race cars better than anyone in the world. They want him to hunt deer, drink liquor and chase women. They want him to be Curtis Turner or Dale Earnhardt, and he's only Jeff Gordon, which by any other standard on earth would be good enough.

Gordon is merely great, and by the standards of those who hate him, that is not enough. How tragic.

But now, 15 races into the season, the point leader is Mayfield, who is almost as young, every bit as handsome, more personable and, at least from the available evidence (we in the overreacting press declared him a superstar just Sunday), absolutely unflappable.

"Jeremy is a nice guy, fun to be with," said his car owner, the thoughtful Michael Kranefuss. "Since I first met him, he hasn't changed one bit. His ability is almost deceiving because he is so friendly. He is living proof that you don't have to be grumpy to be a great race driver. You can be a nice guy."

Mayfield admitted that finishing near the front, without winning, was getting old. He said he was growing weary of the the constant questions: "When are you gonna WIN?" Yet he recognized there was nothing unreasonable about the questions. "It was a source of motivation more than anything else," he said.

The annoying consistency was not what Mayfield wanted.

"Man, we were top-fiving them to death," he said. "That levels off. We wanted to be at that next level."

"Anybody who thinks this is about money is dead wrong," he added. "I would have paid money to win this race. That's what we do. We pay the price to win. I don't race for money. Jeff Gordon doesn't race for money, and neither does Dale Earnhardt.

"I want to win races and that (Winston Cup) trophy."

Kranefuss remembered a test session when Mayfield grazed the wall, then, on the way into the pits, placed several small pebbles in his mouth. When crew chief Paul Andrews leaned into the window, Mayfield spat the pebbles out in his hand. Andrews freaked.

It's a joke, Paul. A joke.

"He hasn't changed yet," said Kranefuss, "and I am absolutely convinced he is not going to change."

LEE MORSE   *He Never Had a Chance*
august 1997

Lee Morse was unable to perform a job for which he was never really prepared.

Morse, once a prominent figure in Ford Special Vehicle Operations and a prime mover in Ford's Winston Cup program, stepped down this week from his position as general manager of Geoff Bodine's team. He played a prominent, if unwitting, role in one of the season's bonafide disasters.

A year ago this week, Bodine won the Bud at the Glen, the annual road race held only half an hour or so from Bodine's hometown of Chemung, New York. Since he purchased the team of the late Alan Kulwicki in late 1993, Bodine has won a race here and there, but consistency has been a problem. The 48-year-old driver took bold steps to rectify that problem during the most recent off-season.

First he hired Morse, who had one of the biggest rolodexes in the sport. The title was general manager, but the job was to raise enough money to pay for Bodine's expansive plans.

Then Bodine, with the encouragement of Morse, negotiated a consulting agreement with drag-racing legend Bob Glidden to work on innovative ways to squeeze additional power from Bodine's Ford racing engines. The word on the street was that Glidden and his son Bill were bringing an entirely new perspective to engine building.

The entire motorsports world stood up to take notice of the Bodine initiative. And it remained standing as these best-laid plans fell apart.

This is, of course, simplistic, but the point must be made in retrospect that drag racers—Glidden earned his success in the National Hot Rod Association's Pro Stock division—are accustomed to racing for a quarter of a mile, not 500 miles. Did the Glidden-derived tuning of Bodine's engines have anything to do with the rash of engine failures the team suffered? The official word is no.

The DNFs and crashes mounted. Bodine needed even more money than his sponsorship with QVC raised. By mid-season, Bodine was close to the end of his financial rope. The expected associate sponsorships Morse was charged with raising never materialized. Bodine was forced to sell off a goodly share of his team to two new investors whose names have not yet been announced.

Morse, for his part, stepped aside quietly, intimating that there soon would be no place for him in the new operation. The businessmen who will take over the team reportedly have sons who will be charged with much of the management of the team.

Like most of the Ford performance hierarchy, Morse is primarily an engineer, not a salesman. Contacts or no, he probably never had a chance.

JERRY NADEAU *With a Little Help from a Friend*
march 1998

Who is Jerry Nadeau? From where did he come? How did he draw the attention of Bill Elliott and Dan Marino when they decided to form a new Winston Cup team?

For several years press releases have detailed the latest exploits of the young Connecticut driver. He was the first American to find success in racing in the Opel Cup, a European series. The releases originated in the offices of Cotter Communications, an eminent Charlotte public-relations firm.

Just another young hot shot with a rich daddy? Amazingly, no.

Nadeau's father is a roofer. He arrives in the Winston Cup as the long shot in a distinguished rookie class. But, like Kenny Irwin and Steve Park, he will have the benefit of decent equipment and solid sponsorship.

Several years ago, Nadeau placed a call to Tom Cotter, asking for advice in forwarding his career. Cotter is a former public-relations director of Charlotte Motor Speedway, and his firm has handled the careers of Darrell Waltrip and Bobby Labonte, among others. Something about the voice at the other end of the phone impressed Cotter, who told Nadeau to meet him the following morning.

What Cotter didn't know was that Nadeau was calling from Connecticut. Without offering complaint or explanation, Nadeau hopped in his car and drove all night to Charlotte, where he made the 9 a.m. appointment.

"When I found out that Jerry had driven all night to make that appointment, yes, it impressed me," said Cotter.

Cotter took on Nadeau as a client "pro bono," a term normally used to describe the practice by lawyers of donating part of their time to worthwhile causes. Literally, the Latin term means, "for good." More specifically, the arrangement meant that Cotter represented Nadeau without being paid. A young member of the Cotter team, Melanie Cannon, took on the Nadeau "account" and wrote releases for him. She became just as impressed as her boss.

Using his considerable influence, Cotter persuaded Richard Jackson to put Nadeau in his Pontiac for several Winston Cup races. Though on the surface his performance seemed undistinguished, Nadeau was good enough to draw Elliott's attention. Anxious to lay a foundation for the future, the veteran chose Nadeau as the driver when he finalized plans for a second team. It is the opportunity of a lifetime. With the backing of FirstPlus Financial, Nadeau will drive an aqua-and-orange Ford Taurus with the colors and No. 13 of Marino, the most prolific passer in the history of the National Football League.

"I am convinced that Jerry Nadeau can achieve whatever it is that he sets out to do," said Cotter. "Fortunately for our sport, this is what he wants. I think he has the talent, which is the most important thing, but also the dedication and the ability to be a great spokesman for his sponsor and the entire sport."

Nadeau is fully aware of his own Cinderella story.

The new team was announced last November at an Atlanta press conference complete with live band, spotlights and a big-screen satellite hookup with Marino, who answered questions from his Miami home.

"I am so thankful for all the people who have put me in a position to realize my dreams," said Nadeau. "What has happened didn't come about because of me. It came about because of all the people who, for some reason, wanted to do something to advance my career."

Cotter is putting the finishing touches on his role in this Walter Mitty story by organizing a Jerry Nadeau Fan Club. The team's official PR representative is now DMF Communications, which also handles the McDonald's account of Elliott's own No. 94 team. The first team is located in a new Statesville, North Carolina, shop. Nadeau will operate out of the shop Elliott still maintains in Dawsonville, Georgia.

Will Nadeau bring this incredible tale to a suitably romantic ending by blossoming into NASCAR stardom? He is a decided underdog in the rookie-of-the-year race, a less proven commodity by far than Irwin, a former open-wheeled champion who won two superspeedway races last year in the Craftsman Truck Series, or Park, who won three times as a Busch Grand National rookie driving a Chevrolet owned by Dale and Teresa Earnhardt.

"I have no doubt that he will succeed," said Cotter.

Nadeau seems to generate that conviction in everyone he meets.

COTTON OWENS *Still Plying His Trade*
june 1998

SPARTANBURG, S.C.— Don't think that, just because Cotton Owens hasn't been around lately, he doesn't know his racing.

At age 74, the hall-of-fame driver, mechanic and car owner still participates in the weekly Saturday night wars. Owens, remarkably spry, turns the wrenches for grandson Kyle Davis' weekly outings on the dirt of Gaffney's Cherokee Speedway. In recent years, he has maintained as many as three cars for Kyle, brother Brandon and first cousin Ryan Owens. Ryan, a graduate of the University of South Carolina, now lives in Columbia, South Carolina, where he works for a public-relations firm. Brandon, who graduated from Erskine College with a perfect 4.0 grade-point average, is enrolled in medical school in Charleston.

That leaves Cotton and Kyle.

"I think the best drivers come off of dirt because you can get a car completely out of control, fight it and save it," says Cotton. "On up to the superspeedways, a driver who got started on dirt will never give up even though he's lost a car. You still drive it, even if your chances aren't one in a hundred, you wrestle with it. The drivers who do that came off dirt."

As a driver, Owens won only nine Grand National (now Winston Cup) races, but he may have been the best modified driver ever. He won more than 100 races during the 1950s alone. As a car owner, he hired only the best, including David Pearson, Junior Johnson, Fireball Roberts, Buddy Baker, Marvin Panch and Ralph Earnhardt.

The elder Earnhardt, who died of a sudden heart attack in 1973, was a lot like Owens, a short-track master who never got much of a chance in the big time. That was the destiny of Ralph's son, Dale.

"Ralph was an awful easygoing person. We were great friends," recalled Owens. "I got out of Modified and into Grand National, and I helped him a great deal on shock absorbers for his Sportsman car. He was like myself, a mechanic and a driver, and he had to look after his family first. He was great at that. As far as a person, I don't think you could find a finer person than Ralph. In 1961, I ran him at Charlotte and Daytona; we ran 3 or 4 races together. I felt like he deserved the chance."

In another footnote, it was actually Owens who gave Mario Andretti his first taste of NASCAR racing.

"We threw a deal together, through Bill France and Ray Nichels, to furnish a car for Mario at the July 4 (Daytona) race," said Owens. "He was one of the greatest. He was smart. He hadn't driven any Grand National cars, and the first thing he said was, 'We need to go to dinner and talk about this situation.' We went out to dinner, and he asked me every aspect about how to get accustomed to that race track. The engine blew up when he was running second, and at the time he was running down the lead car with less than 100 miles to go. He done one heck of a job. Of course, he came back the following year and won [the 1967 Daytona 500]."

No visit with Owens is complete without a few David Pearson stories. Cotton considers his fellow Spartanburg resident the greatest driver of all time, but Pearson was not so great that Owens couldn't put him in his place on one very memorable occasion at the Richmond Fairgrounds.

It was September 14, 1964, and supposedly Everett (Cotton) Owens was retired as a driver. Pearson was 29. Owens was 40.

"David was young and wasn't listening to what we had to say about getting in and out of the pits," Owens remembered. "He'd come flying in the pits so fast that he'd run the crew back over the wall, and we were losing time on every pit stop. I got so mad, I told him, 'There's gonna come a day when I'm gonna show you, 'cause you won't listen', so I decided I'd take another car to Richmond and run in the race myself.

"I outqualified him by about seven or eight positions. I was inside on the second row. I went and won the race, and it liked to killed him [Pearson ran second]. He thought the crew had turned against him or something or another, but it was all in the learning process. It's very hard right now for a rookie to come in and know all the secrets of getting in and out without losing too many positions. He was upset, but he began to listen to us and win races, more so than give them away."

ADAM PETTY *The Same Bright Star Still Shines*
october 1998

CONCORD, N.C.— There's something you need to know about the Petty family.

They are incredible personalities, every one of them: the stern, unrepentant Lee; the proud, smiling Richard; the irreverent Kyle; and now, ladies and gentlemen, boys and girls, make welcome perhaps the most effervescent of them all, 18-year-old Adam, who will soon be coming to race tracks close to your hometown.

The Pettys have won 263 Winston Cup races—for those of you keeping a scorecard at home, the tally is Lee 55, Richard 200, Kyle 8. But as great as they are as drivers, they could just as easily have formed a dynasty of radio/TV personalities.

Richard Petty wrote the book on fan and media relations. He passed on his communications skills to Kyle, who, if he were winning races, would be dominating the column inches and air time in every major market.

The mission of Adam, like the crew of the starship *Enterprise*, is apparently to go where no man has gone before.

Adam Petty is 18 years old, tall, skinny and using Clearasil, and he could sub for David Letterman right now.

On Wednesday night at Charlotte Motor Speedway, he won the very first Automobile Racing Club of America (ARCA) race in which he ever competed. By family standards, this is nothing new. Kyle was also 18 when he won the ARCA 200 at Daytona in 1979.

After Adam's victory, all the kid needed was a guy sitting nearby with a snare drum.

Of his dad's victorious ARCA debut, Adam said, "He told me that when he won his, he was out of control. Was he 18?

Same age as me. He was an 18-year-old kid, and they turned him loose going 200 miles an hour. What else is an 18-year-old kid gonna do? He's gonna 'mat it' and go."

Something else needs to be said about the Pettys. They have their own language, one that is quickly emulated by other racers. Richard began calling other human beings "them cats" 35 years ago, when there were people known as "beatniks" using such terms; he and all the other Pettys still use the phrase. In Pettyese, "mat it" means take your foot and shove that accelerator to the floormat, something Pettys and them other cats are fond of doing.

Adam's general description of his dad Kyle was, "My dad's a big kid. All he is is a kid in a man's body. We're best friends. I can go to him with anything. He can come to me with anything."

Continuing his devastating description of his prominent, unconventional family, Adam said, "My dad and I are really close. My grandfather Richard came from the old school, the school of hard knocks, where you let a kid learn on his own. I think my dad saw that and wanted to do things a little different. I can't say enough about my dad. He's had hard things in his racing life, but he's always been there for me."

He had The King nailed when he noted, "He gave me a hug in victory lane. That was the first time I've ever got a hug from him. He said I done a good job, but we gonna talk about how much I scared him over there on the back straight."

Apparently, the EasyCare 100 was supposed to have been a nice little learning experience for the royal family, generation four. On the radio, Adam was repeatedly told not to take chances, to just "ride." In 1979, legend has it that Richard was shouting in his son Kyle's ear all day long, yet Kyle somehow managed to win the race anyway.

At Charlotte, Adam noted, father Kyle "probably stayed off the radio because my grandfather got on his nerves so bad back in 1979."

But this fun-loving youngster, a kid by his own admission, had to push that damned envelope; funny thing is, he won his first ARCA race, just like Daddy.

"I had to run him (Bobby Hamilton Jr.) down and pass him on the outside coming out of turn four," said Adam. "My daddy and my granddaddy told me not to do that here.

"I think everybody was in shock 'cause I did it. They kept saying, 'ride, ride, ride' and I kept thinking I been riding for about 15 laps now. When you got a Wallace (Mike, who finished second) coming, you gotta go."

Adam Petty's life, alas, is not all fun and games. He was devastated just this past Labor Day, when confusion on pit road caused him inadvertently to run over (and kill) his ASA team's crew chief, Chris Bradley.

The tragedy was eerily similar to what happened to his grandfather in 1965. Chrysler Corporation, then as closely identified with Lee and Richard Petty as Santa Claus with Christmas, had pulled its teams out of the sport in a rules squabble with NASCAR. Richard went drag racing, and his out-of-control Plymouth Barracuda skidded into a spectator area, killing an 8-year-old child.

"The one who made me feel at home first was my grandfather, Richard," said Adam. "He went through the same thing . . . [It] devastated him, but it pretty much made him a better person, too.

"My mind's at ease a little bit more . . . A chaplain told me, 'This can make you or it can break you.' I think it made me. I had to grow up. I'm not gonna say I've become a man, but I've grown up. Racing's my life. I've spent 12 years of my life racing, ever since I was six. If I tried to quit right now, I'd be 12 years behind because I'm not no good at nothing else."

Adam Petty is a young man in a hurry, full of youthful exuberance and fated for greatness by the accident of his birth.

Of the $21,250 he won at CMS, he said, "I didn't see none of it. I'm building a house, and it's going into that. How 'bout that? My mom'll tell you. I'm building this house. I'm an ASA driver and an ARCA driver right now, but the way she's building it, I'm a 12-year Winston Cup veteran. I ain't never gonna make no money. I'm gonna be in debt the rest of my life."

What next?

"You never know. What happened tonight might lead into

something for Saturday," said Adam, dreaming of Charlotte's Busch Grand National race. Apparently, those dreams won't come true, but he'll be at Talladega and St. Louis later this fall, and maybe he'll be driving regularly in NASCAR's Triple-A minor leagues next year.

It is your destiny, young Skywalker.

LEE PETTY *Forgotten Legend*
december 1998

In November 1959, Ned Jarrett's career as one of NASCAR's elite drivers was just beginning. Substituting for another another driver, Jarrett won a 100-miler at the Columbia, South Carolina, dirt track. In the process, he ran afoul of the great Lee Petty.

After receiving his trophy and a check for $800, Jarrett, then just 27, encountered the fierce patriarch of the Petty dynasty.

"Boy, what makes you think you got the right to shove me out of the way?" growled Petty.

Jarrett replied that he thought Lee had been blocking him, and that when he finally pulled alongside, he had felt justified in doing what he had to do to take the lead.

Three days later, Lee arrived at North Wilkesboro, where Jarrett had no car to drive.

"Why didn't you tell me you needed a ride?" Lee, suddenly friendly, asked. "I'd have brung you a car up from our shop."

Jarrett recalls the incident fondly: "I'd had to stand up to him to earn his respect. What he said to me that day earned mine."

On the surface, it is virtually impossible to believe that Lee and Richard Petty are father and son.

Within the realm of American motorsport, Richard is known simply as "The King". Only part of his renown is derived from the fact that he won nearly twice as many NASCAR races as anyone else. Even now, more than six years after he drove his final race and 14-and-a-half years after his last victory, Richard's sunny personality and coast-to-coast smile illuminate the Winston Cup Series. He is the star that warms this small solar system.

Richard Petty became the most successful driver in NASCAR history in the late summer of 1967 when he surpassed the 55 victories recorded by his father between 1949 and 1961. The son went on to win 200 times, perhaps granting more interviews and signing more autographs than any athlete anywhere. He was the first stock car racer to become a household name, and during his long, wildly successful career, seldom did anyone root against him. They may have wanted some other driver to win, but they never wanted Richard to lose.

In terms of work ethic and commitment to excellence, Richard learned at the foot of the master, his old man. But Lee Petty never signed many autographs. Few had the courage to ask.

The founder of NASCAR's royal dynasty—his grandson Kyle and great-grandson Adam are still racing—will turn 85 in March. Lee remains relatively active and is an astonishingly proficient golfer. But the man who helped lay the foundation for what has become America's fastest growing sport will no longer have anything to do with it.

He won't even say why.

His grandson, Kyle, recalled: "Let me tell you a story about my granddaddy. When they inducted him into the International Motorsports Hall of Fame, they called him and asked him to come to Alabama for the induction ceremonies. He told them he wasn't interested. So they called back a week later and offered to pay his expenses. He told them he wasn't interested. A week later they called him a third time and offered to pay him $1,000 on top of his expenses.

"Granddaddy finally told them, 'Look, I got plenty of damn money, you understand that! I ain't interested.' That was that. He didn't go."

Last year the R. J. Reynolds Tobacco Company sent a photographer up to the family compound in Level Cross, North Carolina, for a painstakingly arranged photograph of the four generations of racing Pettys. Lee refused to have his picture taken. He was mowing grass and didn't have time. Richard and Kyle merely shrugged their shoulders. They knew better than to intervene. The photographer, who has been shooting races for decades, chased the old man on the riding mower around for a while and finally flagged him down. Lee could not be persuaded. The picture that was taken contained only three generations.

Last fall Adam Petty, Lee's great-grandson, won an Automobile Racing Club of America event at Charlotte Motor Speedway. When asked what Lee had thought of the teenager's victory, Richard replied, "My father was interested in one generation, and that was his. Fourth generation? He probably don't even know this is happening."

A recent request for an interview was greeted first with the assertion that Lee made it a practice not to review his career on the telephone. When the reporter offered to drive up to Level Cross that very minute, or to come at any available time, Lee said, "I'm not interested," and hung up.

Not surprisingly, the 50th anniversary of NASCAR was celebrated with very little mention of Lee Petty, the first three-time champion, still seventh all-time in victories, who sired its greatest driver and one of its legendary mechanics (Richard's brother, Maurice). Omitting Lee Petty from the history of NASCAR is akin to recalling baseball without Ty Cobb. But Lee Petty's anonymity is by his own choice. When NASCAR unveiled its 50 greatest drivers in a glittering Daytona Beach press conference last February, Lee didn't show up.

Some say old Lee evolved into the troll under the bridge because of the way his career ended. He sailed out of Daytona Beach International Speedway in 1961, two years after winning the first 500, and spent months in a hospital. After a few more attempts behind the wheel, he gave it up and left the glory to Richard.

Others say he harbors a bitterness toward NASCAR dating back to a decades-old belief that the France family cheated him out of at least one more championship.

Maybe Lee just got tired of trying to relate to journalists. In his book *Fast as White Lightning*, Kim Chapin recalled one of Petty's abrupt interviews:

"In the midst of one such dialogue with another reporter several years before, Richard Petty's father, Lee, the grizzled patriarch of the most successful family in Grand National history, suddenly asked his pesky questioner, 'What sport do you play?'

"'Tennis.'

"'Then,' said the old man, walking away, 'there's no point to me even talking to you.'"

Some say he is just a crotchety old coot and that the only thing that has changed is that, when he was young, he was a crotchety young coot. The drivers from his era who survive remember Lee without affection. He was considered a mean, ruthless competitor who would do anything to win. Lee even filed a protest after the race in which his son took his first checkered flag. Lee won the protest, and Richard had to wait. The son wasn't surprised. The first time Richard ever raced on the same track as Lee, the old man put the future King in the wall.

The legendary mechanic Smokey Yunick was frequently at odds with Petty. Not surprisingly, Yunick had few nice words when he was interviewed for Peter Golenbock's book, *American Zoom:*

"I never got close to Lee Petty. . . . There wasn't too many people who liked Lee Petty. Petty was a good driver. Herb Thomas and him tried to get along, but Petty was a two-faced, dirty driver, and I would find it real hard for him to scrape up too many friends in racing today. He didn't start driving until he was 39 (actually 35). Lee had the same trouble Herb Thomas did. He was born very poor, had had a terrible struggle, started and did very well in race cars, worked like a dog and was a very good driver, though he wasn't the caliber of Curtis (Turner) or Tim (Flock) by any means. But survival has much to do with speed, maybe more so, and Lee was a survivor."

When Richard was too young to even think about driving a race car, he did his best to help his dad on race days. Once, during a pit stop, Richard crawled up on the hood to wipe his father's windshield and made the mistake of lingering after the tires had been changed. Lee stormed back out onto the track with Richard hanging on for dear life. He took a complete lap before pulling back into the pits so that his son could leap to safety. Still angry, he reportedly administered a profane, public tongue-lashing to his son afterward.

"When I was a boy, the basic team was Daddy," recalled Richard. "He worked on the cars, did all the work, towed the car to the race track and drove it home after it was done with."

Another time Lee installed wing nuts and armor on the sides of his Oldsmobile, shredding his opponents' cars in the same fashion as the famous chariot race in *Ben-Hur*.

Lee never forgot a slight and seldom let a feud end quietly. During the 1950s, Pure Oil Company had a promotion in which it guaranteed any engine, even a racing engine, from failure due to oil breakdown. Whether because he never read the agreement, or because it had been incompletely presented to him by a friendly salesman, Petty took that to mean Pure would replace any engine he blew. When they refused to do so, he switched to Texaco and boycotted what was then NASCAR's official supplier for the rest of his career.

"When he stopped racing, that was it," said Kyle. "He decided he wasn't going to have anything to do with it, and it was time for the rest of the family to carry on."

Few men have ever abided so completely by a personal vow.

A few years ago, before the annual Darlington pre-race golf outing, someone walked by Lee Petty's table and asked, "Lee, did you ever get that trophy back from Beauchamp?"

Lee had won the first Daytona 500, in 1959, only after NASCAR originally awarded the victory to a relatively obscure Iowa driver named Johnny Beauchamp. It took several days before photographic evidence was found to prove Petty had crossed the finish line ahead of Beauchamp.

Petty looked up, smiled—on those rare occasions that he

uses it, he has the same grin of his son, grandson and great-grandson—and remarked, "No, I reckon that (sonuvagun) Beauchamp took it to hell with him."

Then he paused with the timing of a comedian.

"I mean, uh, to heaven."

Lee Petty raised his magnificent family during the Depression, when surviving was all a man could aspire to. He is by no means a bad man. He just learned life's lessons from an Old Testament God.

RICHARD PETTY *The King*
september 1998

DARLINGTON, S.C.— This year the Southern 500 weekend banquet was something special, yet it was an event as predictable as the onset of May flowers after April showers.

Richard Lee Petty joined the Stock Car Hall of Fame.

As overwhelming as his career was in terms of success, Petty most deserves his place in the National Motorsports Press Association's hall for his personality. He was NASCAR's first household name. Other drivers may have matched his virtuosity: David Pearson, Dale Earnhardt, Curtis Turner, Fireball Roberts and Junior Johnson are among those who may have matched Petty in terms of talent.

No one has ever matched The King's charisma, at least not among the grease-stained gladiators of American motorsports.

Everyone knows that Petty won 200 races, nearly twice as many as any other driver. Less well known is the fact that Petty, he of the gaudy cowboy hat, the wrap-around sunglasses and the Ultra-Brite smile, may have signed more autographs than any man who ever lived.

"When that stuff first started, there wasn't but three or four thousand people at some of the races," Petty recalled. "When the boys were loading up the car, I'd sit back there and sign autographs.

"When you stop and sign somebody's autograph, it means

you're giving something back to the people. I'd watch those people's faces, and I saw they were genuinely thrilled with it."

Petty won 27 races in one year, including 10 in a row. No one else in history ever won more than 18. Petty even owns the modern record of 13 victories in 1975. His effect on the history of stock car racing is comparable to that of Babe Ruth in baseball, only Petty's records still stand.

Gordon wins races. Earnhardt did so for many years, as did Darrell Waltrip. All three have been rewarded for their excellence by being vilified by the fans.

It never happened to Petty, who was admired even by those who did not want him to win. When asked about this phenomenon, The King humbly notes that, during his greatest years, he drove Plymouths and Dodges in a world in which most fans were fans of either Fords or Chevys. As a result, Petty was neutral. The Chevy fans hated the Fords, and the Ford fans hated the Chevys. If Petty won in his electric-blue Plymouth, they could live with it.

Nice try.

The truth is that Petty was and is beloved because he is simply one of the nicest, most pleasant, most patient human beings who ever lived. There is no telling how many of his 61 years were spent signing autographs, answering repetitive questions from nosy reporters, posing for photographs and indulging the brainstorms of various public-relations men.

"When I came along, there was a different set of circumstances," he said. "The fans at that time, they didn't think (David) Pearson or (Bobby) Allison, or myself, was anything special. We'd just sit out there with them at the motel. The press has taken the drivers now and put them on a different plateau than the one we were on.

"I think the drivers now would prefer to do it like we did, and just hang out, but there ain't no way. I think nowadays there's more press than there used to be fans."

Given the nature of the sport, it is really not surprising that Gordon, a polite and pleasant young man, gets booed every time he shows his face in public. In auto racing, it is the nature of the beast. Every week 43 cars take the green flag, each car with some measure of fan support. Gordon is the favorite every week, meaning that to the fans of every other driver, he is the threat. Many fans love Gordon, but their cheers can never equal in intensity the catcalls of those who are against him.

What is surprising is that Petty, the most dominant driver of them all, never faced the boos.

That is more impressive than all of his triumphs.

RICKY RUDD *Fighting the Good Fight*
february 1999

DAYTONA BEACH, FLA.— What has Ricky Rudd learned from five seasons of dual life as a driver and car owner?

"What I've learned is that I understand a lot more about what some of the car owners I used to drive for were doing," Rudd said. "It never hurts to walk in the other guy's shoes."

It is fashionable to say that the day of driver-car owners like Ricky Rudd is past. It seems like just yesterday that it was the rage.

Racing is a game of follow-the-leader in more ways than one. Alan Kulwicki's 1992 championship, which seems more like a miracle with each passing year, bred a generation of drivers who thought, by running their own operation, they could duplicate Kulwicki's success.

"It was never about money," Rudd said. "We weren't trying to make a bunch of money. We thought, as racers, that we understood what it took to be winners."

In point of fact, the driver-car owners were doomed almost from the start. College-educated Kulwicki had uncommon business sense. For every success story like his, the years since have been littered by the failures of men like Geoff Bodine and Darrell Waltrip, who ended up selling off their operations after initial successes proved misleading. Even Brett Bodine, who still persists in this two-steps-forward-one-step-back game, has succeeded only to the extent that he has managed to keep the doors open.

Since he went on his own in 1994, Rudd has cost himself trips to victory lane, and he knows it. He became his own boss because he wanted to provide for a future outside of the cockpit. He soon realized the growth of multi-car, well-financed teams would make his own victories few and far between. Yet Rudd has won at least one race in every season, and his overall streak of 16 seasons with a victory is currently the sport's longest.

"It seems like there's always a struggle," Rudd said. "You put a quality team together, and you can't keep it intact, because when you do good, other teams with more money buy off your key personnel. Last year we had a struggle to get the Taurus to run. When I went on my own, politically it was not correct, but race-wise, it was the best thing for my future, and I don't regret the decision."

In the ongoing controversies over NASCAR rules, Rudd must clearly side with those who want to keep expenses under control. But he understands why that side does not often rule the day.

"If you've got the resources behind you, you want to use them," he said, "and when you get right down to it, the leading teams, the ones who have the unlimited budgets, are also the ones with the most clout. I don't know that that's unfair. It's not like I don't have the option of owning a multi-car team. There isn't anything in the rulebook that says I can't do the same things that those people are doing. It's just hard for someone like me to get far enough ahead that you can provide some of those advantages for yourself."

NASCAR racing was not always this way. It's not that the Winston Cup Series of today is inferior to the old days; it's just that it is different.

"Yeah, I could name you four or five things I'd like to see changed," he said. "From my point of view, they're things that could be done to improve the quality of racing without costing you an arm and a leg. But naturally, guys like Rick Hendrick and Jack Roush and Roger Penske don't want to see rules in place that keep them from using the resources they've built up. They don't want to give up the advantages that come with their investments. I can understand that. We all end up doing the best we can do with what we have. Some of us have more than others, and that's the way the world operates."

Rudd remembers his Winston Cup debut, way back in 1975 at Rockingham. He was 19 years old, in the best condition of his life, yet he remembers being physically spent in the closing laps. Donnie Allison, a veteran in his 40s, lapped Rudd late that day.

"He waved thank-you to me for getting out of his way," Rudd said, "and I was just amazed because I looked over there and he was driving with one hand. What I had to learn is that you've got to learn how to relax in a race car. What wears is you out is tensing up.

"What I miss about the old days is the close contact. But these guys today, they're a pretty darn good bunch of drivers, and you don't see people taking the cheap shots like you did a few years back. It used to be that two or three drivers caused all the wrecks. The whole group now is a cleaner bunch to race. Basically back then it was a war. Drivers didn't look at the big picture. They'd be focused on winning that one event they were in and not looking at down the road."

Rudd remembers his Dover victory in 1997, when Mark Martin dogged him in the closing laps.

"That's a good example of what I'm talking about," he said. "At the end that day, Mark was coming hard. He had every chance to cheap-shot me in order to win that race, yet he didn't. I think it's because, when you're all racing each other week to week, and everybody's running for the points, you know it comes back to haunt you and that you drive everybody else like you'd like them to drive you.

"In a way, I miss the way it was back then. But when you look at the big picture, I enjoy racing this group we've got now."

FELIX SABATES   *Don't Stop Him, He's Rollin'*
september 1998

DARLINGTON, S.C.— Owner Felix Sabates may not be winning races, but he consistently delivers the goods in the post-race quotes department.

The excitable Cuban millionaire, saddled with three Chevrolet race teams that are decidedly not the equal of Jeff Gordon's, got carried away, as usual, when describing Gordon's latest extraordinary victory in Sunday's Pepsi Southern 500.

"I think Jeff Gordon is the best driver of all time," said Sabates. "I think they've got the best crew of all time. If he had been driving a Ford this year, he might be undefeated. I think he's the best driver who's ever driven a stock car."

So frustrating is Gordon's domination of this sport, even to others in the grand, exalted brotherhood of Chevrolets, that soon Sabates got so heated that his comments dissolved into Stengelese, no, maybe even the murky depths of Yogi Berra.

"Those guys (presumably Ford drivers) are crying about tires and all that," said Sabates. "They ought to go to a pawn shop and get a bugle, and they should go home tonight and blow their own horns against a mirror and watch the reflection in the mirror rather than complain about soaked tires."

What the heck does that mean? It gets better.

"Jeff Burton is a pretty damn good race-car driver," rattled the Charlotte businessman, "but Jack (Roush) better be careful about what he says because, sometimes, it might come up and bite him in the butt."

"If you're going to blow smoke, he ought to blow it in front of a mirror and let it come back and hit him in the face."

OK.

The point is that Gordo is having a neurotic effect on stock car racing. Each week his competitors work harder and harder, and each week Gordon whips them worse and worse.

Sabates wants to win as badly as Rick Hendrick, Gordon's ailing car owner. In 1992, Hendrick went looking for a fresh face and found Jeff Gordon. Four years later, Sabates took a similar tack and found another Gordon, Robby (no relation).

Jeff Gordon evolved into a two-time champion, a force of nature who became the sport's dominant competitor. Robby Gordon's one season in the Winston Cup Series (1997) was an utter disaster.

Now Team Sabco employs the triumvirate of Sterling Marlin, Joe Nemechek and Jeff Green. While Sabates and his minions continue to huff and puff and aspire to greatness—Marlin, for instance, had a creditable eighth-place finish Sunday, a mere two laps down to Gordon—no one in sports is going to equate "Marlin to Nemechek to Green" with "Tinker to Evers to Chance."

To paraphrase a country songwriter, Hendrick got the gold mine, and Sabates got the shaft.

MIKE SKINNER *Lone Wolf*
february 1999

DAYTONA BEACH, FLA.— OK, so Jeff Gordon won the Daytona 500, and he wasn't the most popular driver on the race track. Gordon is accustomed to fending for himself.

Perhaps in time Mike Skinner will earn the respect and trust of others in the Daytona International Speedway draft. For now, though, Skinner will seethe in the knowledge that his Chevrolet was as fast as Gordon's and Dale Earnhardt's, and the best he could manage was fourth place.

"Naturally you're a little frustrated," said Skinner, still looking for his first Winston Cup victory. "It's hard to watch people go draft with your opponents instead of your own team, but that's the way it goes. We're at Daytona, man. This is restrictor-plate racing. Nobody is going to help anybody if they've got 31 (Skinner's number) on the door."

Seventh-place finisher Kyle Petty knew he was riding along behind a marked man.

"Poor ol' Skinner, he's going to have to go around and give out thousand-dollar bills," said Petty. "Nobody will go with him. I felt sorry for him. I got in line behind him, and I couldn't help him much."

The central problem for the Skinners of the world is that,

with the cars' horsepower limited by the dreaded restrictor plates, it's tough to get by without some help from friends. For some reason, no one will give Skinner an even break. One of these days, they're going to let him in the club, but for now he must try to be patient and await his turn.

"I thought we could have won the race," said Skinner. "We didn't, and I'm a little disappointed. It could have been a lot worse. I saw some cars leave on a rollback (truck that carts off wrecked cars), so it turned out pretty good."

Actually it could have been worse. Skinner could have been Tony Stewart, the rookie who started on the front row but never had a chance. He hung in there the best he could, but the engine in his Pontiac eventually gave out. Stewart was eighth after 10 laps, ninth after 20, fifth after 30 and 10th after 60, but he never showed up again. It was an exercise in frustration, one that Stewart probably expected after Earnhardt labeled him "a mirror-driving son of a gun" in Thursday's qualifying race.

For guys like Skinner and Stewart, Rockingham, the site of next Sunday's race, must be a welcome place to get back to normal. Stewart missed winning a Busch Grand National race by inches last year. Skinner has never fared as well on tracks like Rockingham as he has at Daytona and Talladega, where his cars have always run well. Still, he won't have to be nearly as politically correct there.

"This is our third year," said Skinner, rookie of the year in 1997. "We should start winning races pretty soon."

JIMMY SPENCER *He Always Puts on a Show*
october 1998

TALLADEGA, ALA.— As Jimmy Spencer stood in the shadow of his team's rig, he looked like exactly what he was: a race driver. He stood there, smiling, answering questions, accepting congratulations, all the while sweating in his red-and-white driver's outfit.

Spencer did not win the Winston 500. He merely put on the

best show. If the race had been decided by a panel of judges holding up numbers for the quality of his moves and the degree of difficulty, Spencer would have won the race by a landslide.

He would have won even if the judges had been Russians, or worse even to the driver of the Winston-sponsored No. 23 Taurus, non-smokers.

But, to paraphrase no less an observer than Johnny Cash, life ain't easy for a boy nicknamed "Mr. Excitement."

Jimmy darted side to side, he put his Ford down on the apron, and he squeezed it up against the concrete wall. He spent the whole day doing what Dale Earnhardt used to call "stirring things up." Spencer was the straw that stirred the drink. Hell, he was the Waring blender that frapped the shake. In the end, he had to settle for fourth place.

Few of his detractors have ever met Spencer. If they had, they would not be his detractors. At age 41, he remains the closest thing this generation has seen to the late, beloved DeWayne "Tiny" Lund. Spencer is affable, fun-loving, chubby and mischievous. He has all the characteristics that used to make race drivers famous. He fits the current generation poorly, but if he had been able to race the Curtis Turners and the Junior Johnsons, he would probably be in the hall of fame right now.

Jimmy likes to play, and in the perilous Talladega Superspeedway draft, too many of his playmates won't.

"You gotta have the best car," said Spencer afterward. "D. J. (Dale Jarrett) had the best motor. All I could do was put myself in the right position to be there at the end, roll the dice and hope for the best.

"I'm just glad I'm racing against a bunch of professionals. Bobby Labonte and Terry Labonte and Dale Jarrett, every one of them. They pushed and shoved as far as they could, but they never went over the line. . . . I hope they respect me for what I did (Sunday). . . . Naturally, I wanted to win, but I just came up short."

Spencer won a race here, Daytona too, but that was four long years ago. He drives the only car Travis Carter owns, and at Talladega, a man could use a teammate or three. Truth be known, Talladega races are often decided by politics, and Spencer ain't no politician.

"Nobody can take anything away from the run I had today," said Spencer. "Sometimes I could have used a little help, but you gotta try to win. I don't give a damn what anybody says. When it comes down to the end, you gotta look out for yourself, and you can't get mad when the other guy does the same thing.

"What do I regret? Nothing. Well, maybe I could've done one thing different. I should've had a Robert Yates motor. Them last few laps were nerve-wracking, I'll tell you. Wow! I'd be sixth, fourth, third, second . . . then I'd blink my eye, and I'd be sixth again."

BILLY STANDRIDGE *The Sport Needs Little Guys, Too*
august 1997

BROOKLYN, MICH.— He blends in rather nicely with the garage-area scenery, exchanging greetings, smiling, swapping stories and standing around in the shadows of the huge race rigs that mobilize the Winston Cup Series.

But Billy Standridge is no fan, tourist or flack. He is a race-car driver without a ride, and as such, he feels the need to be seen. Out of sight, out of mind, and Standridge wants to be one of the names that arise whenever there is a vacant cockpit.

Standridge, from Shelby, North Carolina, has been on the sideline for about a month, when he was released as driver of the No. 78 Ford fielded by Jim Wilson. The argument could be made that Standridge got more out of the car than anyone else has, but in the end his performance was not good enough. A couple of drivers have succeeded Standridge with no particular success. Gary Bradberry will start 36th in the car in today's DeVilbiss 400.

"It's pretty tough to go to the tracks, but you've got to do it," said Standridge, a 44-year-old veteran of many a feature on the short tracks of the Carolinas. "It's the only place you're gonna find a ride, that and staying on the phone."

Standridge has never had what most would consider a fair shot. At the Winston Cup level, he has never climbed into a car with financial backing remotely equal to that of the competitive teams. At this point in his career, the big-time opportunity is unlikely to occur. Standridge is preparing to fend for himself, and it will not be the first time.

"I'm working on a few things, maybe for next year," Standridge said. "We're building a car for Talladega. We started doing that as soon as I got released from the 78 car."

In an age in which stock car racing is ruled by multimillionaires, it is tough for a journeyman driver to compete. In fact, it is tough for a journeyman driver to make the field. But racing is what Standridge does.

"Before I went with the 78 team, the 47 car was mine," said Standridge, using numbers to keep from having to recite miscellaneous items like makes and colors and sponsors. "Buck Johnson, he was in with me, he owned 20 percent of the team, but I owned the rest of it. I managed the team, built the cars. I know how to put a team together. I was pretty much driver and crew chief and everything else, so I know how to set up a car.

"When I went to the 78 team, that was what I was hired to do—be the crew chief. Then I became the team manager, and then I became the driver. We got the team up and running, too."

It is a two-steps-forward, one-step-back life for Standridge, and that's when he's lucky. He'd have to live in a castle to be able to put all the lessons up on the wall.

"I'm not getting into a situation again where I'm just hoping for a miracle," he said. "I wouldn't go back to Talladega if I wasn't convinced that we had a good motor, a good chassis and good people to help me with the car."

The NASCAR life may not have been a good life, but it has been Billy Standridge's life.

TONY STEWART *On the Charts with a Bullet*
february 1999

DAYTONA BEACH, FLA.— It's time we all got to know Tony Stewart, because we are probably going to be following his exploits for a long, long time.

Stewart represents the latest wave of youthful Hoosier hotshots to roll into the Charlotte area seeking fame and glory. First there was Jeff Gordon, then Kenny Irwin, the 1998 rookie of the year, and now Stewart, who lists Columbus, Indiana, as home.

At one time, all three drove United States Auto Club (USAC) sprinters for a legendary Indianapolis car builder named Bob East. East calls his creations "Beasts," taking the first letter of his first name and pairing it with his last. It is a fitting title for one of those open-wheeled monsters that prowl the short tracks of the Midwest, and it is a fitting title for the devil-may-care chargers who eventually are climbing out of East's Beasts and becoming NASCAR beasts themselves.

Stewart and Gordon will share the front row—Gordon, naturally, managed to secure the top spot—when NASCAR's greatest race gets underway on February 14. As surprising as this may seem, it will be the first time the two, both 27 years old, have ever raced side by side. Way back in 1994, the two were in the starting field of the same Hoosier Dome midget race, but both crashed before they ever got close to each other.

Gordon is the unchallenged master of the NASCAR domain. Stewart has bumped around a few more years getting here, but he has never failed to impress. He became the first driver ever to win three USAC titles (Sprint, Midget, Silver Crown) in the same year. He started his first Indianapolis 500 on the pole, and he won the championship of the Indy Racing League in 1997.

In short, it may take Stewart a while to get the hang of these 3,400-pound stock cars, but he is unlikely to dawdle for long. In fact, Irwin had better get his helmet strapped on straight in a hurry if he wants to keep the title of chief young lion in the Gordon menagerie.

Just as Gordon once did, way back in 1993 or so, Stewart carries with him the aura of greatness.

"It's been surprising how much the drivers, car owners, crew chiefs, everybody have welcomed me to this series," said Stewart. "I really thought they would be more cutthroat about it. The truth is, they've really been really good about making me feel welcome. During testing, Ernie Irvan walked by me, then he stopped, came back and said, 'Hey, I'd really like to welcome you to the Winston Cup Series.' I don't think there was any kind of motive in that. He really wanted to help me out."

What Stewart is going to find out is that success will bring a chill to the atmosphere of conviviality and good humor. So far, the top dogs in the NASCAR kennel are honored that so prominent a young racer has opted to join their ranks. The minute he starts rumbling toward victory lane, that's when he will become a threat to their family's well-being, their ability to put food on the table, and most important of all, their souvenir sales.

Based on his qualifying run, Stewart is rumbling along at a stout pace already.

KENNY WALLACE Mr. Nice Guy
Feels the Heat
april 1997

MARTINSVILLE, VA.— After Kenny Wallace won the pole for the Goody's 500, you couldn't have dragged him out of the press box with two sets of chains and a John Deere tractor.

Long after the formal press conference was over, Wallace could be heard in the back of the room, erupting in laughter as he goodnaturedly took questions from anyone who wanted to talk to him. Several security officers got one-on-one interviews.

Veteran sportwriters exchanged glances and rolled their eyes. "Who is this guy? He *likes* us."

Kenny Wallace, the youngest and most personable of stock

car racing's three Wallaces, likes everybody. He could no more hold a grudge than a folk singer could hold a job at a nuclear power plant.

Yet nothing has come easily for smallish, freckle-faced Kenny Wallace. He could never quite win the Busch Grand National championship, despite several strong bids. He flopped as a Winston Cup rookie and was fired by Felix Sabates after only a year. He drove briefly for Robert Yates while Ernie Irvan was injured, but it was Dale Jarrett who was hired as the full-time replacement.

Through it all, Wallace never showed more than a trace of bitterness. Sabates is "really a good old guy," he says. Jarrett has been "an inspiration to me because a lot of people tried to write him off."

Furthermore, Kenny realizes his funloving personality has on occasion hindered his racing career. Some have said he ought to be at the Comedy Club instead of the Speedway Club.

"Nice guy," noted one mechanic. "What is it they say about nice guys? They finish last."

Mr. Nice Guy is starting today's race first, and he's not mincing words about the need to stay there. Wallace considers the Goody's 500 to be the greatest opportunity of his career. He vows not to relinquish the lead without a fight. The freckle-faced kid is a scrapper. Even if his team doesn't have the resources of a Yates, a Hendrick, a Roush or a Childress. Kenny's car owner is named Martocci. Filbert Martocci. The name sounds like he ought to be peddling frozen fish and canned spaghetti.

"People listen to me laughing and telling jokes, and they read that to mean I don't take it seriously enough," said Wallace. "They don't know me. Nobody is more competitive. I've just got a lot of nervousness inside me, and laughing and telling jokes is the way I burn it off."

All the Wallaces are intense. Rusty dissipates his intensity by being so talkative and frank that his mouth sometimes gets him in trouble. Mike is a deal-maker with an oft-cited knack for the business end of the sport.

Kenny Wallace qualified .213 of a second faster than Rusty and .380 better than Mike. But Rusty has won the spring race at Martinsville four years in a row, and Kenny has never won a Winston Cup race anywhere. If you polled the grandstands today, not one in a hundred would forecast a Kenny Wallace victory. Probably close to half of them would name Rusty as the favorite, even though he's starting 15th.

Kenny could be a comedian. And like another comedian, he don't get no respect.

RUSTY WALLACE *It Gets to You After a While*
october 1998

AVONDALE, ARIZ.— It nearly drove excitable Rusty Wallace crazy. He was having a good season. His team was undeniably better. But he hadn't won.

No one—not Jeff Gordon, not Dale Earnhardt, not Dale Jarrett—enjoys the spoils of victory more than Russell William Wallace. As much as he paid lip service to what was happening to him—"It's bound to happen," "We've just got to hang in there, and wait for things to fall our way"—it was gnawing at him. All the comments were just rationalizations. He had to drive out on the asphalt and do it.

Was he concerned? As concerned as a mother with her kid still out after midnight. He was discontented and brooding, so much so that, when the long-awaited victory finally came, he could hardly enjoy it without looking ahead to greater conquests in the weeks ahead.

"Yeah, I was concerned," Wallace admitted finally. I ended last year with 47 victories, and my main goal this year was to win my 50th because of NASCAR's 50th anniversary. A lot of special plans have been made for this 50th victory. I'm really looking forward to it, and I do have a shot at it now."

He certainly left himself no room for error. To get his 50th during NASCAR's 50th, Wallace will have to win at Rockingham next week and then at Atlanta. He'll have to go 2-for-2, this from a man who had been oh-for-59 since a Richmond victory early in 1997.

That's Rusty. Earlier this year he made the bold statement that he had won 25 races in the state of Virginia, failing to note that, to come up with that figure, he would have to count six wins at Bristol, which is in Tennessee; three at Dover, which is in Delaware; three at North Wilkesboro; and one of his two at Charlotte, the last two tracks being in North Carolina.

They're close to Virginia. Tennessee and North Carolina border Virginia. Delaware almost does. What's the problem? And what's to keep Wallace from taking three in a row to end the season?

"I'm going to two tracks that are really good to me," said Wallace. "Rockingham? I've won a ton of races there, and I'm taking the same car, a car that we nicknamed 'Streaker' (Sunday) because we kept the streak alive, so that's the official name of the car now. Its name before that was PR-22, the 22nd car that was built at Penske Racing. I can't tell you, I don't think I've ever had a car that was that dominant that long."

For the record, Wallace has won five races at North Carolina Speedway in Rockingham, and he has won twice at Atlanta Motor Speedway, though both were before the track was rebuilt.

The streak Wallace was talking about is of 13 years' duration. He has won at least one race in every season since 1986. When Wallace made note of the streak after the race, he called it 16 years. Oh, well.

At Phoenix, Wallace was overdue for a victory. Even his detractors would admit that. Was he so overdue that now he is going to win two more in rapid-fire succession?

"I've got all the intentions in the world of doing that," he said brashly. "I'm going into Rockingham with an awful lot of confidence because I tested three days there this year. I started the (February 22) race, led the most laps, thought I had it won and finished second. . . . We had a great car, and we're going to a track I've tested at. Not only did I test there, but I've had a ton of victories at Rockingham."

Once again, Wallace's memory exceeds the facts. At Rockingham earlier this year, Wallace led 74 laps. Mark Martin led 104, the same number as winner Gordon.

"There are all kinds of reasons why I should win Rocking-ham, and there are a lot of reasons why I should win Atlanta. I went to Atlanta this year and finished third, ran up front all day long. The car was super strong, and the only thing different is that I've got a stronger car now."

In the first Atlanta race, on March 8, Wallace finished fourth, not third.

But Wallace believes he led the most laps at Rockingham, and he believes he finished third at Atlanta. He also believes he is going to win both those races the second time around.

Mark it down.

DARRELL WALTRIP *He Still Loves Racing*
march 1998

DARLINGTON, S.C.— Darrell Waltrip has been labeled a has-been. The grandstands are rife with fans suggesting, even demanding, that he give up his ambition for a few more shin-ing moments of Winston Cup glory. Other critics declare that he talks too much.

It would be difficult, however, to find anyone who would say D.W. is a bad person. He is not mean, selfish, or bad-inten-tioned.

In fact, Waltrip is a warm, generous man, a sap for a sad story, the kind of guy who can't walk past a Salvation Army vol-unteer ringing a bell at a mall without emptying his pockets.

At the low point in his career—D.W. came to Darlington having made only one starting field the old-fashioned way— he called a press conference to discuss his plan to raise over $100,000 to pay Tim Flock's medical bills. Never mind that he had spent $1.3 million of his own money on his troubled race team. Never mind that it is up for sale. Never mind that this proud three-time champion has become an object of ridicule in many fans' eyes.

Waltrip talked of his affection for Flock, the ex-champion

who is dying of cancer. He talked of his weaknesses as a car owner and about advancing age and about why on earth he doesn't climb out of that race car and put his considerable verbal skills to use on TV.

But mainly he talked about how, adversity and all, he is still such a lucky man.

"Last week I got to thinking about all my problems," Waltrip said, "and then I got to thinking about Tim and Frances Flock, and about their sons, and about what they must be going through. I felt almost embarrassed that, here I was, feeling sorry for myself."

In a weary, crackling voice imported via telephone, Flock himself answered questions. Waltrip's Chevrolet had been decorated in the white and red of Flock's 1955 NASCAR championship.

"This is the most unbelievable thing I've ever seen," said Flock. "I think it's the most beautiful car I've ever seen. I appreciate Darrell and his team taking the time to build the car with my (number) 300 on it.

"Don't tell (Dale) Earnhardt, but we slipped a Chrysler hemi in that thing overnight."

Someone tried to turn the announcement into a political discussion, suggesting that NASCAR had been negligent in not providing for the retirement needs of ex-heroes like Flock, Herb Thomas and Rex White.

"(Drawing attention to their plight) is not my intention," said Waltrip. "My intention is to help my friend. There are people in this sport who ought to have enough money to help the have-nots, if they need it. I'm not interested in politics; I'm interested in helping a friend who is in need."

MICHAEL WALTRIP   *No More Excuses*
february 1999

ROCKINGHAM, N.C.— More than any other Winston Cup driver, Michael Waltrip is haunted by victory, or more accurately, by the absence of one.

Waltrip, whose brother Darrell is tied for third all-time with 84 visits to victory lane, is a big fat zero in 395 starts. His spirits rose only marginally last week at Daytona, where he finished fifth. It was his first top-five finish since Talladega in April 1996.

"It'll take a win," he said. "I started this season before the Daytona 500 thinking this could be my last season. I've got a great team, a great sponsor, and if I don't run well, I might not be able to get a quality ride next year.

"I gotta win. I'll never feel like I'm accepted, like I'm solid or stable, until that happens."

Waltrip, who moved into a new ride this year, went through a stage of bitterness in which he resented the constant questions stemming from his inability to come through. It seemed to some as though his 1996 victory in the Winston all-star race—an incredible upset but still unofficial as far as the record books are concerned—had been a hindrance, not a help, to his career. Like his brother, Michael has a wonderful way with words, the instincts of a standup comic and a personality eminently capable of weathering hard times. It seemed a shame to see such talents go to waste.

Now, he realizes, the questions are "part of it. There's things people go through in their job every day. You could look at it as a hassle. You folks (the press) have a job to do. You have to ask questions. I'm gonna hear it, and I'm OK with that.

"In the early 1990s, I almost won a couple of races and sat on a couple of poles (three, actually). I really thought I was on the verge. It got away from me; it didn't happen. I went from close to kind of getting away from it. Now I feel like I'm close again, and that's an exciting thing. When I started with the Wood Brothers in 1996, I felt the same way, and at the start, we did do good. For some reason, we let that slip away. I'm back up. I'm pumped again."

Waltrip's latest career shuffle brought him to Jim Mattei's team, located in the same shop and derived from the team the late Alan Kulwicki used to win the Winston Cup title in 1992. This year, Mattei switched from Ford Tauruses to Chevrolet Monte Carlos, a fact that makes Waltrip's impressive performance at Daytona even more so.

Why?

"It's my belief that NASCAR will make those two race cars equal," Waltrip said. "They're going to do what they've got to do to make the competition even. What you have to do is hopefully get one or the other (Ford or Chevrolet) to commit to you. With the Ford situation, they (Mattei) didn't really have any support. Chevrolet is just really committed to our team, with [Team] Sabco [with which Mattei has a cooperative arrangement], wind-tunnel testing, on-track testing, we're a part of that. The most important part for a team to do is get with whichever manufacturer that will help you the most. You just have to be with one that is going to help you the most."

Maybe, just maybe, this is finally Waltrip's year. "I have the confidence that I need to win," he said. "It's just putting all the pieces of the puzzle together. It doesn't matter how good you are and what you know. It's a matter of getting with a team to where we all think a lot."

He was asked what would happen if, once again, things didn't work out, and he found himself without a ride.

"Shoot myself, possibly," he said, kidding. "I'd run Busch, trucks, I'd be racing something. I don't plan on letting that (losing his ride) happen. I don't like to look at it like it'll be OK. It won't be OK. I've got to win."

HUMPY WHEELER *The Visionary as Traditionalist*

january 1999

CONCORD, N.C.— The annual Winston Cup Media Tour has hardly started, and already I've uncovered the most significant fact of the entire week.

Charlotte Motor Speedway president H. A. "Humpy" Wheeler hates cellular telephones.

It's all I need to know. Here is a man who envisions a day, 20 years from now, when races will be held all over the globe and where fans who tune in on TV will be able to race with their favorite drivers by putting their home entertainment systems in virtual-reality mode. Yet he hates cell phones. That alone is enough to make me like him. I myself—and I am ashamed to admit this—own a cell phone.

It all started in November when my rental car had a blowout in Seattle, Washington. There I was, changing a tire in the driving rain, scared that some criminal was going to emerge from the shadows behind a dark warehouse to attack me and my niece. I vowed that, if I got back home alive and with money, I was going to buy one of those godforsaken phones. I don't like them. I think they contribute to the rampant decline of civilization. Yet, because of concerns for security, I bit the bullet and bought one. Humpy is a businessman of international renown, and he doesn't own one. Hurray for him.

Humpy Wheeler may not be the most imaginative man on the stock car racing scene, but he does the best job of letting his dreams be known. His little quirk about cell phones actually tells a lot about his character. It is rare to find a man with his attributes. He has been an important participant in a process that has taken NASCAR racing to bigger and not always better things. Yet Wheeler got his start as a short-track promoter. Gastonia was once the center of his universe. His humble roots are becoming more and more an insignificant footnote in the story of his career, but Wheeler remembers them. Deep down, I think, they are where his sympathy lies.

Singlehandedly, Wheeler enlivened a symposium held Monday by the Cabarrus County Chamber of Commerce on the topic of motorsport. Had he not been there, the group would have consisted mainly of pompous people talking about how important they were. One of the speakers who preceded Wheeler made the following fervent assertion: "Without the concept of NASCAR, NASCAR would not exist."

The members of the audience sat rapt, looking as if they were hearing the Sermon on the Mount. The nonsensical nature of that remark sailed right past them.

Since I was wearing a jean jacket and cowboy boots, I looked pretty out of place in this audience of suits, I mean, as if my being out of place were not already obvious from the smart-alec remarks I kept making under my breath. Jerry Gappens of the speedway walked up and said I looked like Billy Jack. I told him I felt like Billy Jack.

A guy from Hendrick Motorsports proudly declared to the audience that the operation was expanding onto land it had leased from "the late Harry Hyde." Didn't he mean it was leasing land that had once belonged to Hyde? How did it arrive at this agreement with the late Harry Hyde? Was a seance held?

Believe me, I could go on and on. If it weren't for the set of mint-condition Piedmont Boll Weevils bubble-gum cards that were in my press kit, I probably would.

Anyway, Humpy saved the day.

He didn't talk about what Cabarrus County could do to preserve motorsports. Let's face it, the only way the county is going to lose motorsports is if someone blows up Charlotte Motor Speedway, right?

Wheeler said that, in 2020, what is now known as the Winston Cup Series will consist of 30 events, 20 in the United States and 10 overseas. He described a track in northern Germany—he said the name, which was a cross between "wiener-schnitzel" and "neidemeyer"—with a roof over the grandstands, pit area and track surface. In the infield, where no one would be in a condition to notice, they would get wet. He envisioned a race between "global cars" representing six different manufacturers. The cars, he estimated, would average 230–240 mph, occasionally bouncing off liquid-filled protective walls.

By 2020, he said, television would drive the entire industry, as if that's not already true. He said viewers around the world would watch on ultra-high-definition TV's three-dimensional pictures that would make it look like they could walk right into the screen. If viewers wanted, they could push a button and find themselves right on the track in their own virtual-reality-generated cars, dicing with the real drivers, who of course would be unable to notice they were there. All 40 of the cars would be fitted with cameras; others would hover in the air. Back home, viewers could punch up practically anything related to the race, from play-by-play to the physical condition of each driver.

All the drivers, Wheeler said, will have started racing when they were 7 or 8 years old. Few will be older than 40. By 2020, Wheeler said, medical scientists will have discovered a way of identifying what makes a person adept at driving race cars, and those with the gift will be given a chance to receive training almost from the day they are born.

Scary? Yes. By then, the whole world will be scary. We have no idea into what brave, new world we are headed. Or at least, most of us don't.

A view from the stands at the MBNA Gold 400, Dover Downs International Speedway, Dover, Delaware, September 1999. *Tyson Cartwright*

PLACES

ATLANTA MOTOR SPEEDWAY
november 1997

HAMPTON, GA.— The track's name is Atlanta Motor Speedway, but it is not, repeat not, in Atlanta.

It is in Hampton, which is about the same as McDonough, Jonesboro and Griffin, only they've got shopping centers. The members of Hampton's chamber of commerce were ecstatic when they attracted a convenience store to town.

But I like that. I'm a small-town boy, and I like it when I'm on the road and I can stop in at a family-owned store where the old man at the cash register can make change without a computer telling him how to do it. I like the small towns where people know how to bag groceries—they don't slam a 10-pound bag of scratch feed on top of a loaf of bread, and they still use paper instead of plastic.

So when I came down to Atlanta, AKA Hot-lanta, the nightlife capital of the New South, I avoided it like an offshore hurricane. I evacuated the interstate. Drove through the country, where I could admire the lay of the land. Admired a young woman with a pair of towheaded, obstinate young'uns. Stopped in at a shack (the word is not used loosely) called Jackson's Barbecue Corner in beautiful downtown Union Point. Washed down some "Q," Brunswick stew and slaw with a Diet Coke (real crushed ice!) and eavesdropped on the construction workers at the next table talking about what a babe the stop-n-go sign woman was.

In the middle of Shady Dale, a woman sat on the tailgate of a pickup selling boiled peanuts, which she scooped steaming out of a pot, along with fresh, deep-fried pork rinds. I didn't stop, just sat there at the four-way stop sign, watching her shivering in a light jacket as steam swirled all around her.

The fans do not come from Atlanta. There is little tie-in between the Atlanta Hawks and the NAPA 500. The National Basketball Association being on strike will not sell one extra ticket at Atlanta Motor Speedway. Why? Because people who drive city streets, with the possible exception of the devil-may-care drivers of taxicabs, do not develop the great American love affair with the automobile. That love affair is necessary for one to become a stock car racing fan.

When the ear-rattling roar of a V-8 splits the air, a race fan's heart jumps. Anyone who wishes they would do something about the noise does not belong in the stands. Get some ear plugs. Don't spoil it for the thousands who are willfully putting into place an environment that will leave them stuffing hearing aids into their ears at age 45.

I'm one of them. I love that roar. Better to fill my ears with sound while still young enough than never to hear the righteous roar.

The city is full of gadflies who spend their lives worrying about cholesterol, global warming, El Niño (or La Niña, this year's designer weather pattern), secondary smoke and television violence.

Country folk don't worry about that stuff. Life is fatal. Why worry about the things that are slowly killing us? Seize the moment. Come on out to the speedway where we can all revel in the joy of dying very slowly and watching cars go fast.

BRISTOL MOTOR SPEEDWAY
august 1998

BRISTOL, TENN.— Drive by Bristol Motor Speedway early in the evening, after the sun has dropped below the crest of the surrounding mountains, and you'll think the Mother Ship has landed.

Bruton Smith has installed eerie blue lighting in the bowels of his crater-shaped race track, and if E. T. has not been sighted, he (she? it?) certainly could not be far away.

What Smith and his onsite managers, Jeff Byrd and Wayne Estes, have done to this rugged mountain oval is stunning. One hundred and thirty-four thousand seats now ring the high-banked concrete.

To fulfill his plans, the Charlotte tycoon erased a mountain . . . a big mountain that once bordered the back-stretch stands.

The Egyptians would have passed on this project. "Nah," they would have said, "we'll just stick with pyramids."

Romans would have stuck with minor architectural projects like the Colosseum. No doubt Caesar Augustus would have taken a look at Smith's plans, laughed and said, "We Romans prefer to stick with building that which is possible."

Then the proud Emperor would have lapsed into his native Latin, leaned over to the youthful Smith (2,000 years his junior) and said, "Et tu, Bruton?"

Sorry. Couldn't resist.

Bruton Smith is a man who does not take "no" for an answer, which is why he regularly achieves what is considered impossible. It is entirely possible that a day of reckoning will come when he or his heirs race past the rational bounds of logic, sense and credit. Until then, I cannot help but look at Bristol Motor Speedway and admire the can-do spirit that departed most of the country's business community in the giddy decade after World War II.

Smith has stacked up a mountain of seats to replace what once was a mountain of nature. He has taken north Texas prairie and built the second largest stadium in the United States.

n to being loony as a spaniel on he-
p truck had apparently wasted away
s in Margaritaville. Without remem-
away a 12-pack of Molotov cocktails
his Nissan, our hero decided he
d timeout, briefly quieting the cadre
had driven and flown to the scene,
When he ignited his cig, the home-
causing Smokin' Joe to take leave of
senses.
ind you.
n then threatened to jump off the
had enough sense to realize that a
od below might cause him to linger
re going to the Great Asylum in the
esperate—I mean, c'mon, Dilbert,
rday Night Live skit—our antihero
s brains out with his trusty shotgun.
lis entire head flew off, right there on
V).
recommends for the kiddies.
do with stock car racing, or sports, or
ts in China? Bear with me.
rnia, and this race track, that on
h up close and personal as Cousin
the family, foisted his sad, pathetic
9-inch diagonal. On Thursday, how-
e sporting event about which I was
—first-round qualifying for the Cali-

t that gave me the rare opportunity
t served like a volleyball did not have
ng event.
le, or at least that's what they tell me.
e day was orchestrated by escapees

Milk and Honey . . . I do love L.A.

DARLINGTON RACEWAY

september 1997

DARLINGTON, S.C.— What Darlington Raceway brings to stock car racing is not anything that is going to impress a Wall Street banker or a television advertiser. But if a day ever comes in which the Southern 500 is not a vibrant part of the NASCAR landscape, then the Almighty would have every excuse to stop anybody named France or Smith or Penske, or anybody descended from them, at the gates of heaven.

Darlington is not a raceway; it is a shrine. And in this sleepy town, the raceway's aura is a way of life.

On Saturday morning, they were preparing for their annual parade in downtown Darlington. In the parking lot of the Piggly Wiggly, men were sitting in the backs of their GMC pickups, spitting tobacco, shelling boiled peanuts and peeling off a greenback or two so that the young'uns could buy a balloon or a goofy hat. The old buildings were suddenly alive with the chatter of families and the laughter of children. At the Darlington Outlet Store and Sam's Pawn, bunting was being strung up to enliven the faded bricks.

Further out Harry Byrd Highway, fans were lined up to visit the Joe Weatherly Stock Car Museum, now located behind the back stretch of the 47-year-old track. A parking spot at the Raceway Grill was particularly hard to come by since most of the space was taken up by the souvenir trailers in the gravel yard.

The lore of Darlington is as splashed with rustic stories as it is with driving skill. For every time that Fred Lorenzen and Curtis Turner scraped sheet-metal lap after lap, there is a tale about a driver chasing a sportswriter through the garage area or about a faux pas by the public-address (P. A.) announcer.

For many years, the late Ray Melton was the Darlington P. A. man, and he ruled his kingdom with an imperial, albeit homespun, air. One year, at a race sponsored by a nearby automobile dealer, a line of identical Jeep Cherokee Scouts paraded around the track during the pre-race ceremonies.

During those days, a popular Darlington promotion was to

ondos for hundreds of thousands of dol-
of his speedways and entertained others
way dining.
me Smith's creme de la creme. It took de-
ven control Charlotte Motor Speedway,
its finances. Texas has been plagued by
gerous track conditions. Atlanta is strain-
the boss's plans for it.
ly love this place. The level of enthusiasm
g. They turn the surrounding hills into a
of Boy Scouts, swarm over the access
f ants marching into battle, and fill every
ng Oklahomans.
been 40,000 out here watching the Win-
ce Saturday evening. Practice!
boro, Smith ranks between communists
al opinion polls. But in east Tennessee, he
t of mountain folk very happy.

PEEDWAY

— On Wednesday, I was standing on the
ooking the California Speedway garage
something unusual. At the gates, where
the garage and onto pit road, there were
lace. Obviously this was done as a crowd-
t most tracks accomplish this with a few
rly policemen.
ck into the press tower, which overlooks
the live story of a nutcase who parked his
d threatened to start shooting at people
ed at his health maintenance organiza-

ommissioning the Mormon Tabernacle
ises of my HMO, but I'm not going to go
n about it, either.

invite Boy and Cub Scouts from around the state of South Carolina to attend at cut-rate prices. Many years the back stretch was thickly populated with the green and blue uniforms of the scouts.

Melton received a call in the P. A. booth. "Be sure to mention those Cherokee Scouts," a track official said.

"Righto," said Melton. "I'll get right on it."

He then turned on his microphone and earnestly intoned, "Ladies and gentlemen, we'd like to call your attention to the back stretch at Darlington Raceway, where today we have in attendance thousands of Cherokee Scouts from all over the great state of South Carolina."

Darlington once had its own pagoda-type scoring stand, just like Indianapolis. Its rickety press stand, long since replaced, was known as "Balmer's Box," because of the time that a driver named Earl Balmer sailed over the wall in turn one and sent writers scurrying for cover.

But the reason that Darlington should always have a place on the NASCAR schedule has nothing to do with its small-town charm, its rich history or its nearly a half-century of tall tales and legends. Darlington should remain because it is, quite clearly, the greatest test of stock car racing skill. With the lone exception of Larry Frank way back in 1962, the Southern 500 has never been won by a driver with less than legendary skills. Look at the multiple winners list: Cale Yarborough 5, Bobby Allison 4, Dale Earnhardt 3, Buck Baker 3, Herb Thomas 3, David Pearson 3, Bill Elliott 3, Fireball Roberts 2, Harry Gant 2 and Jeff Gordon 2.

When seeking immortality on these hallowed grounds, the novice need not apply.

DAYTONA INTERNATIONAL SPEEDWAY
july 1998

DAYTONA BEACH, FLA.— If you come back to Daytona in October for the running of the Pepsi 400, and if the prices are high, for once don't be too hard on the local merchants.

Like much of the outlying area, the local economy is toast due to forest fires.

On the Fourth of July, there were no fireworks in Daytona Beach. The streets were nearly vacant. The switchboards at beachfront motels were jammed. Hundreds of tourists were calling in to cancel their reservations. Bad news sure spreads fast.

At Louise's Pizza Restaurant, the cute blond waitress was terminally depressed.

"This area is dead," she said. "It was tough enough around here already. This town is barely hanging on."

Little joy existed in the few motel rooms that were occupied. Mostly, there were a few lingering race fans, men with Earnhardt caps drinking beer out of tall brown bottles as they watched their kids playing in the pool, and refugees from Ormond Beach and Flagler County, where nearly 100,000 people had been evacuated. Firefighters, heretofore housed in college dormitories, moved into the motels, where rates were slashed. A room that went for $150 on race week could be had for $35 tops. As long as the smoke stays inland, it ain't half bad.

Among the year-round residents, the old were blaming the young, and the young were blaming the old. Meanwhile, the mortgage payments were getting harder and harder to meet.

Elderly residents are intent on keeping the rowdiness away, which means that Daytona Beach has become an unwelcoming venue for college kids on spring break. If the senior citizens could rid the world of race fans and bikers, each of whom pump what is tantamount to solid gold into the motels, restaurants and bars, they would gladly do so.

"And they all vote," said the waitress ruefully.

"Believe it or not, the race fans and the bikers don't cause many problems," said the owner of the restaurant, whose name was not Lucille. "The bikers may drive up and down Atlantic Boulevard all night, but they don't tear stuff up. They spend more money in the community than the NASCAR people, just because they spend most of their time on the beach, not out at the speedway."

The wildfires have turned world-famous Daytona Beach

into a ghost town, and the next bad thing that is going to happen is when the fires end. Then the firefighters will return to their homes in Wyoming, North Carolina and everywhere else. The evacuees will return to Flagler and Ormond, where many of them will breathe sighs of relief to find they still have homes.

On the oceanfront there is no danger, not unless the great fire Godzilla learns how to swim the Halifax River or sweep across the bridges. But the panic on the coast is fueled by dollars and cents, and the next time money is going to sweep into Daytona will be in October, when NASCAR returns.

DOVER DOWNS INTERNATIONAL SPEEDWAY
june 1998

DOVER, DEL.— I'm warming on Dover.

They shortened the races to 400 miles, ending that tired old "24 Hours of Dover" tag we used to put on the races here when they were 500 miles (and 500 laps) around the concrete mile.

I guess I'm getting a soft spot for the whole state because racing is so important here. They crammed about 125,000 people around and inside of what used to be a horse track Sunday.

Let's think about that number. It's more than three times the number that lives here. It's a quarter of the entire state's population. It's worth $30 million to the local economy. Proportionally, that's like letting the state of Texas loose in, say, Germany.

And we know what that would cause. Bedlam. Anarchy. A shortage of chicken-fried steak.

That Delaware is even a state almost defies belief. Memorialized because it was the first in the Union, Delaware is almost completely enclosed by the Atlantic Ocean, certainly an estimable body of water, and Maryland. Maryland!

A lot of Delaware is like everywhere else used to be. There are towns with names like Pearson Corners and Marydel, and they have fruit stands on darn near every corner. An explosion in the infield of the race track last Thursday somehow managed to knock out the cable in my hotel for three days. Honest, that's what they told me when I called the front desk. So, on Sunday morning before coming to the track, I actually learned how Martha Stewart makes homemade mayonnaise. And the best pecan praline cookies . . . hush your mouth. I mean, *Sportscenter* was unavailable.

The race fans are friendly. Before Sunday's race, there must have been a thousand of them standing around on the banked front stretch, scratching their heads and wondering how they could get in the garage area for some nifty autographs. Even the ones wearing Earnhardt T-shirts seemed pleasant enough. I didn't see any of them beating the stuffing out of a clean-shaven Gordon fan. They were bearded and tattooed, yes, but civil.

OK, the races are long, the weather is hot, and the Italian sausages in the press box cause massive . . . well, never mind. But it's also the only place on the circuit where one can combat that gastric, er, uh, pain with free milk and apple juice.

I had a wonderful time Sunday between trips to the can.

Let me count all the other things I adore about Dover.

The Atlantic Book Warehouse is a great place to pick up great works of literature—you know, like, *The World's Greatest, Hot-do-mighty, Fried Green Tomatoes Recipe Book*—at reasonable prices. Within a mile of the track is a shopping mall, a slot-machine casino, and a barbecue restaurant called "Where Pigs Fly."

I kid you not.

Last September, a Busch race at Dover was being held on the same afternoon that the Delaware State Hornets were playing a football game across the street. A Dover Busch race draws perhaps 65,000 fans. A Hornets home game, judging from the size of the stadium, draws perhaps 3,000.

Can you imagine the confusion on the part of the loyal Delaware State alumni as they run into traffic in, say, Annapolis?

"Great day, Inez, I had no idea A&T brought so many fans on the road."

This whole state is about as wide as the distance from Gastonia to Shelby, North Carolina, which means you're never far from seafood. Seafood and I, we got a good thing going. I like the way these folks up here spread out the paper, dump crabs out of a big pot and just start chawing away up to their elbows. Kind of like a pig pickin' back home.

I'm a Dover fan, I tell you. I have seen the error of my ways, when in bygone years I would sneer about the absence of such high-brow luxury items as towels provided clean, fresh and daily. Never more will I nitpick over those towel-less rooms costing $150 a night on race weekends. Besides, I now reside in the Hampton Inn, not the rustic Budget Inn, and at the Hampton I get all the amenities of other places for just $15 more a night . . . even if the propane-tank explosion in the infield did somehow knock a TV satellite out of earth's orbit.

I shall miss Dover when I arrive back in the Carolinas this afternoon. The glistening aluminum of the stands. The enlightened gentility of the State Troopers who directed me into the track. The staccato symphony that is Al Robinson on the P. A.

Heaven on earth. With a V-8's roar.

INDIANAPOLIS MOTOR SPEEDWAY
august 1997

SPEEDWAY, IND.— No, a fan is unable to see the entire track at Indianapolis Motor Speedway.

Trust me on this. It's OK.

TV and radio announcers are fond of referring to Indianapolis as "the greatest spectacle in motor racing." In a profession noted for relentless hype, in this particular instance they are right on the money.

Mr. Spock would get goose bumps watching the start of the Brickyard 400. The speedway has a reported 307,000 seats. Atop the mammoth grandstands are the signal flags of racing: red, checkered, white, black, blue with a yellow stripe, green and yellow. The pre-race ceremonies have been carefully planned as a wholesome celebration of Americana. The Indiana State University marching band, casually decked out in shorts and white T-shirts, has toasted the throng with "Stars and Stripes Forever." The heroes have arrived at their cars after waving to the crowd from the backs of convertibles lapping the 2.5-mile track. The most stirring scene? Fifty-year-old Darrell Waltrip, accompanied by his two little curly-headed girls.

More than three hours later, winner Ricky Rudd tried to come to grips with the intangible allure of the place.

"After the race is over, you ride around in that convertible," said Rudd after acknowledging the gratitude of all the fans. "There is something about having grandstands on both sides of the race track. There are just walls of people.

"You had people who had Jeff Gordon shirts on, Earnhardt shirts, some Rudd shirts. But they are all just race fans, and they are cheering as you ride by. Coming down pit road really meant something special. I think every crew member out there on pit road came by and gave me a 'high five' as I was coming to victory lane. That really means something special when you've got your competitors who want to see you win. I guess we're in somewhat of an underdog role. It's pretty neat."

When NASCAR came to Indy for the first time in 1994, there was some resistance, some snide comments from the traditionalists accustomed to the sporting world's annual focus on Indianapolis during May. There was a vague dread of the Southern interlopers coming to the Midwest to steal some of the thunder.

No more. The stock cars are as respected now as the low-slung, earthbound rockets. The stock cars are nowhere near as fast or maneuverable, but what a show!

What a spectacle.

Even the Indianapolis Police Department has joined the NASCAR fan club.

Defying every rowdy stereotype they had been led to expect, the authorities have learned that the stock car crowds are better behaved than the ones that arrive at the speedway for Memorial Day weekend. Most of the NASCAR fans have come to watch a race. Many of the Indy-car fans just come to party.

"The crowd has been very pleasant," said State Police Sgt. David Schipp, quoted in Sunday's *Indianapolis Star*. "It was a good crowd, and there were some very good people. We had no problems.

"A lot of people come from all over the U.S. because they really love NASCAR racing," he added. "They're not so concerned with partying. As a rule, the Brickyard 400 crowd is much easier to deal with."

Country has come to town, and the city folks are tickled to death.

LAS VEGAS MOTOR SPEEDWAY
march 1998

LAS VEGAS, NEV.— Las Vegas Motor Speedway is a beautifully designed marvel of engineering wizardry rising out of the desert north of Sin City. On its grounds are facilities suitable for virtually any form of racing—NASCAR, CART, IRL, NHRA, SCCA, Pro Sports Car, World of Outlaws, AMA—currently popular in the United States.

Beyond the actual tracks—the 1.5-mile oval, half-mile dirt track, drag strip, .375-mile paved track, two road courses—is enough land to park an allied expeditionary force. As they discovered years ago in Phoenix, the non-arable desert land is fairly easy to come by, and there's no need to pave the lots, because in a normal year it rains all of four inches. Also in the vicinity are Nellis Air Force Base, two different prisons and enough nuclear bombs in storage, somewhere up in the mountains, to turn Earth into the fourth or fifth rock from the sun.

The most surprising aspect of Las Vegas Motor Speedway, as it turns out, is how dignified it looks. It cost $200 million, but it really looks like a speedway that could have been built in Kansas City or Des Moines. The money here was spent on functionality, not glitter, and that is rather unusual for Vegas.

The main grandstands are painted in patriotic stripes of

red, white and blue. The luxury boxes are rather plain-look-ing, architecturally, but there are 14 different sets of elevators ready to hoist celebrities, businessmen and other VIPs to their individually-appointed vantage posts.

Then there is Sin City 15 miles to the south.

It is not enough for the city to have a life-sized replica of the Statue of Liberty staring down on Las Vegas Boulevard. Oh, no, that looked a little plain to the powers that be, so they added a life-sized model of the Sphinx behind it, and then a black-glass recreation of an Egyptian pyramid.

One hotel has an entire amusement park—inside—in the lobby. Caesar's Palace looks like the United States Supreme Court Building, stretched up to 30 stories. The Rio is a high-rise hotel striped in purple, green and orange. These mon-strosities assault the senses at the same time the casinos within assault the pocketbook.

The Mirage . . . Circus Circus . . . New York New York . . . The Orleans . . . Bally's . . . The Sahara . . . all fitted with gigan-tic fountains, scantily-clad dancing girls, enough neon to ring Saturn and the omnipresent "ch-ching" of big fortunes turn-ing into small ones.

It amuses me when restaurants advertise their chocolate desserts and label them "decadent." Mississippi mud pie is not decadent. It is fattening. Las Vegas is decadent.

MARTINSVILLE SPEEDWAY
april 1998

MARTINSVILLE, VA.— There are those among us who do not think the Winston Cup Series should come to places like Martinsville Speedway. These big-city boys think no venue smaller than Las Vegas should host the rapidly booming uni-verse of Bill France.

Then again, some of us like the mix.

On Saturday, before the Featherlite Modified Tour race got under way, the preacher launched into a spirited, evangelical prayer inviting the 25,000 or so in attendance to give their hearts to the Lord.

One of the writers in the press box sighed and said, "Oh, boy, another non-denominational prayer."

One of his neighbors opined, "There's something you don't understand about places like this in the South. Everybody's name is Smith, Jones and Davis, some of us are black and some of us are white, but all of us love Jesus."

This opinion was underscored later at Frith's Barbecue, where the waitress set a group of diners straight before she even took their orders.

"I love a hairy-chested man," she said. "O' course, he's got to be saved."

The superstars of the *MECW* were on display at Patrick County High School in Stuart, half an hour's drive away. The MECW is a regional federation of pro rasslers (wrestlers being those who grapple amateurly). Opinion was divided on what the letters stood for. "Mildly Entertaining Championship Wrestling" was the most apt of the suggestions rendered by the professional wiseacres at the gym. "Mighty Educational Championship Wrestling" was another.

The most intriguing character was the rassler who had, it was intimated, escaped from a mental institution. In fact, he wore green scrubs with the words "Bell View Hospital" lettered in magic marker on the back. Facially, he bore an uncanny resemblance to Confederate general Stonewall Jackson. Several times he paused to ruminate by sitting in the corner atop the ropes, flashing his otherworldly grin and polling the crowd on whether he should disarm his opponent by breaking his limbs or strangling him. When the opponent trapped him, the mental patient, rather than have his head bashed against the turnbuckles, removed the man's hands and obligingly bashed his own head against the turnbuckles. He was crazy, see. Get it?

The local football players, all wearing T-shirts proclaiming their recent district championship (with the slogan, "Those who stay will be champions") and sitting together, taunted all the rasslers as the matches ran deep into the night. Among the combatants were two midgets (one of whom was fond of biting the referee on the behind), four women and only three (not counting the women) who weighed as much as 200 pounds.

The following obituary appeared in the local paper on Sunday morning:

"On Friday, April 17, 1998, at 6:25 a.m., Thomas Hunter Bell Jr., 76, set sail for his last voyage.

"He was a sailor, a salesman, a gambler and the best steak griller in the world. . . . "

Thus immersed in the lore of the region, it occurred to me on Sunday as rain fell on the speedway that racing is probably too staid, too respectable for this colorful region. A combination race-car driver/moonshine runner might work, as once it did, but today's drivers do not have the versatility.

If one wanted to rule this region, if he wanted to become a folk hero along the lines of a Junior Johnson or a Charlie "Choo Choo" Justice, it would take a truly versatile person in this day and age.

The combination of a rassler and a television evangelist would work. Imagine the pre-bout interview between the godly rassler, let's call him Ernest Angry, and his opponent, a godless communist smuggled in from Soviet Georgia (somewhere below Macon, methinks).

The Reverend Angry could beat the poor Georgian to a pulp, raise him from the dead and have him give his testimony, all for the price of one ringside ticket.

This could work.

MICHIGAN SPEEDWAY
june 1997

BROOKLYN, MICH.— Roger Penske did not build Michigan Speedway, so the mundane quality of stock car racing is not specifically his fault. That awful legacy has been attributed to a long-forgotten man named Larry LoPatin.

LoPatin even built an evil twin, Texas World Speedway, that went belly up.

Penske, who bought the track at a bargain price when it was headed for financial ruin, knows a good investment when he sees it.

He just doesn't know what a real race track is.

Sometimes the rich and powerful become so insulated that their concept of reality changes. Penske, the multimillionaire industrialist and world-renowned motorsports entrepreneur, somehow came up with the mistaken notion that Michigan Speedway represented some sort of classic framework of motorsports competition.

Penske cloned Michigan and built California Speedway with the same dimensions.

There, too, the racing usually stinks.

The Miller Lite 400 turned into sort of a tag-team boredom match between Mark Martin and Jeff Gordon. It will be interesting to see the TV ratings. This might be the first stock car race that ever lost viewership to a golf tournament. Or a game show. Or a bass tournament.

Usually a breathtaking afternoon at Michigan offers the spectacle of cars running out of gas in the waning laps. This time we did not even get the proverbial "mileage run." There have been great races here; the most recent was in 1991, when Dale Jarrett defeated Davey Allison by inches at the line. Usually, though, either it's a runaway or "The leader's out of gas in turn four!"

Officials here hide behind the basic statistics. For example, there were 15 lead changes in Sunday's race. Please note: 10 of them occurred because one car pitted and the one behind it stayed on the track. Basically, this was a race between Jeff Gordon, who was fastest for, oh, about 150 of the 200 laps, and Mark Martin, who swept past Gordon at the end. The day's excitement consisted of a four-lap segment (147–150) in which Martin worked his way past Gordon and a three-lap segment (190–192) in which Jarrett relegated Gordon to third. The other race leaders—polesitter Ward Burton, Bill Elliott, Jimmy Spencer, even DARRELL WALTRIP, for gosh sakes

—were just window dressing in an otherwise barren show-room.

The Penske braintrust will also note that, however boring these affairs are, they somehow manage to draw everyone in the Midwest who can't get Red Wings tickets. As usual, on Sunday, the place was jammed.

Here is the secret: Penske has a captive audience. The only other venue within a gazillion miles is Indianapolis Motor Speedway. If this track had to battle for the public dollar with a Charlotte, a Rockingham or a Talladega, inside of six months the garage area would be converted into a flea market.

NEW HAMPSHIRE INTERNATIONAL SPEEDWAY
july 1997

LOUDON, N.H.— The sky above New Hampshire is glorious. Thousands, even millions, of stars shimmer brightly, the heavens undimmed by nearby city lights.

So why, in the state whose motto is "Live Free or Die," is NASCAR here? New Hampshire International Speedway is near Concord and Manchester. By the standards of this tiny state, these cities are as populous as it gets; by the standards of NASCAR and its desire to reach the so-called "major markets," however, they are rather modest.

The answer lies slightly over an hour to the south.

Boston. The metropolis of New England. The city of *The Globe*, which boasts that its sports section is America's best.

The Globe quite rightly takes pride in its sports section, where giants like Bob Ryan, Peter Gammons and Dan Shaughnessy reside. But *The Globe* is only the state of the art if one happens to be a fan of the Red Sox, Celtics, Bruins or Patriots. The only way NASCAR can even guarantee that *The Globe* will even show up at a Winston Cup race is to hold one, and now two, each year at Bob Bahre's frontier track in the lush, rolling hills.

The Globe ran a story about Darrell Waltrip this week. The paper called Darrell Waltrip "the Silver Fox." Race fans know that nickname to belong to the retired legend David Pearson. Country-music fans once hung that moniker on the late Charlie Rich. No one except *The Globe* has ever known D.W. as "the Silver Fox."

Within New Hampshire, however, the speedway is king. The Loudon track represents the only major-league venue of any sort in the Granite State. Fifty or 60 miles from the track, every convenience store, every bar and grill, every barber shop hangs its banner out: "Welcome Race Fans."

The metropolitan newspapers of Boston, Hartford and Providence have never given racing much more than the standings in agate and a couple of graphs of wire copy. Yet the sport is vibrant here.

The Featherlite Modified Tour and the Busch North Series thrive despite receiving only the appreciation of trade weeklies like *Speedway Scene* and *Area Auto Racing News.* They may not care about the short tracks in Boston and Hartford, but the hardworking men and women of the villages plunk down their hard-earned greenbacks at the bullrings with names like Thunder Valley, Thompson and Stafford.

The modified drivers have nicknames like Spittoon, Black Top and, yes, Sweet Breath. *The Boston Globe* is not the type of rag to dub its heroes with those kinds of names. *The Globe* prefers to devote its copy to Red Sox Nation and its Olde Towne Team.

These Yankees are intensely loyal.

Ricky Craven, the only Winston Cup driver from Newburgh, Maine, said: "It's probably the only time this year that I'll have greater applause or greater introduction support than Dale Earnhardt or Jeff Gordon. That's kind of significant for all of us on the Bud (his sponsor) team. I've had a lot of success at New Hampshire, for whatever reason.

"I think part of it is because it's home."

Rent yourself a room on the edge of Lake Winnipesaukee. Get yourself a table at one of the rustic inns and order lobster, or perhaps a bowl of "chowdah." Try cherrystones on the half shell. Pull off the interstate and visit the rest area, which is also the site of a state-owned liquor store. Where else can you find a llama farm or a "temperance tavern"?

The bright lights here are not in the big city. They're in the sky.

NORTH CAROLINA SPEEDWAY
november 1998

ROCKINGHAM, N.C.— I love North Carolina Speedway.

Why? It's small enough that fans can see everything, but it's large enough to run laps at over 150 mph. It's small enough that there are no restrictor plates, but large enough to accommodate crowds of 75,000 or so.

Its elevated grandstands, coupled with its economical size, make Rockingham the best spectator track in NASCAR. The sight lines are splendid.

A driver can make a difference here, and he can't run wide-open all around the 1.017-mile laps. He can't deplete the car's resources because tire wear is so excessive. A track ought to require a driver to be smart, and Rockingham does.

The general manager, Chris Browning, is pleasant and thoughtful. I'm so impressed with the public-relations director, Kristi Richardson, that I've begun calling her "America's Sweetheart." Roger Penske's operatives have taken over the body of this speedway but not its soul.

The first race I saw here was in 1986. Neil Bonnett won it. When the early 1990s arrived and I began covering motorsports more and more, I started calling races here "the 24 Hours of Rockingham." I can no longer make such disparaging remarks. The difference between 500 and 400 miles at a track like this one is dramatic. Fans resisted the shortening of the races, but I can't imagine any of them who still resent the move. The races have become markedly better since they were shortened. Do you want time or entertainment? Most fans, I think, would opt for the latter.

It was at Rockingham that I met country singer John Anderson, sidling awkwardly about in the garage area like the small-town boy that once he was.

Some of my colleagues wish all the races were in places like Las Vegas and Los Angeles. Not me. I like the mix. When I go to Rockingham, I enjoy it as much as the glitter of Las Vegas. In fact, I don't even like Vegas. I think it's tacky. I'd take John Anderson over Wayne Newton any day. Besides, I gamble about as well as I play golf, which is to say that I'm not likely to be joining the Senior Tour at age 50. I've got nearly a decade to practice, but no, I don't think it is going to happen. Meanwhile, my dream is to go to Vegas one year and not lose a dime. Alas, this is about as possible as becoming a good golfer.

In Vegas, there are show girls, neon signs and swarthy guys with pinstriped suits and open-collar shirts. In Rockingham, there are high-school football games, piney woods, sand hills and barbecue. Twenty miles up U.S. 1, there is, well, golf. Golf on courses where they didn't have to move mountains. The Pinehurst/Southern Pines area is as rich in golfing lore as the links of Scotland, and courses are built according to the lay of the land.

Besides, driving to Rockingham is relaxing. Take along some music. Pick out a few country storytellers—an old Tom T. Hall album or something recent by, say, Guy Clark, maybe some 1970s-vintage Merle Haggard—and hit the road. Once the infernal Independence Boulevard is escaped, about halfway to Monroe, the road becomes relaxing. Of course, by that time, the traffic may make you consider a firearms purchase. I recommend an alternate route—up I-85 to Concord, then down through Albemarle, for instance—on the theory that a few extra miles are a small price to pay for peace of mind.

SEARS POINT RACEWAY
may 1997

SONOMA, CALIF— My lack of sophistication is all too apparent in the San Francisco Bay Area. I have obviously never learned to appreciate the finer things in life.

For instance, since arriving here last Wednesday, I have sampled probably five different brands of beer (none to excess, of course). They had names like Sierra Nevada, Anchor

Steam and Friesen. Most of them were fruity. Everything in California is fruity. But that's another story.

At 3Com Park, I bought a cheeseburger, an order of garlic fries (you could smell them 100 yards away), and one of those darn microbrews, altogether value-priced at a mere $16. But they would, however, accept a Visa card for payment. After rushing to the Stick (it was once known as Candlestick Park) in the bottom of the second inning, I leaned back with my $21 ticket to enjoy my $16 meal and realize that my rental car was safely tucked in bed in its $6 space.

Baseball . . . the affordable game!

Halfway through the game, a fan sitting nearby offered me a sample of the marinated venison his father had prepared. It was tasty, I remarked. He asked me to give some to the guy sitting in front of me.

You would have thought I had offered a martini to a minor.

"I don't eat meat," the man said. "It's cruel. Animals shouldn't be slaughtered or hunted."

When the man found it was the meat of deer, he really exploded. I informed him that, while I personally did not hunt deer, I had no problem with it because it kept the animals' soaring population under control. I told him driving through a national forest back home was no less dangerous than tiptoeing through a mine field. The man, who was keeping a scorebook of the game and had called Deion Sanders several vile names, said we should leave Bambi alone.

"How do you feel about plants?" I asked. "They're living things. Do you mow the grass?"

Another potential friend down the drain.

Later that afternoon, a friend and I toured Haight-Ashbury, the famed hippie neighborhood. I bought a Cal Santa Cruz T-shirt because that school's nickname is Banana Slugs. Everything else in the shop was too weird for me. I left my friend, who was trying to find a "conservative" tie-dyed shirt, and went outside to lean against a lamppost and watch a long-haired kid strum his guitar and sing.

The singer was the happiest guy I ever saw. He sang of love. It didn't matter to him whether people listened. He just danced around and strummed that guitar. A blond-headed kid sucked on a Marlboro and started grooving along with him. Two men in front of me started kissing each other. And not on the cheek, either.

My friend walked out of the shop, holding up a T-shirt with Che Guevara or Jerry Garcia or somebody on it.

"You know, I've done some pretty wild, weird things in my life," I told him. "But compared to everybody within five blocks, I'm the All-American boy."

Driving to the track on race morning I made the mistake of changing lanes. A man driving a beige Chevy Citation lagged when the line moved, and I scooted in front of him. Two cars back, a bald-headed guy in a green Ford station wagon went ballistic.

Baldy spewed the most awful of words at me. He was pointing and throwing both arms up. Finally, he sent his poor wife to scold me. She looked embarrassed as I watched her through the rear-view mirror dutifully shuffle up. She leaned into my window and called me a rear end, among other things.

I looked back at her and said, "Ma'am, I go to every race on this circuit, and every week I try to get to the track as fast as possible so's I can get my work done. You tell your husband or whoever he is to take a pill, watch me and take some notes."

Then I handed her an official Sterling Marlin Kodak reporter's notepad and sent her back to Genghis Khan. I think I'm now an advocate of gun control, because if he'd had a pistol, I'd be shot right now.

This part of the country is indescribably beautiful. If the food were any better, it would be against the law to eat. I will not mind visiting again.

But California and I do not fit.

TALLADEGA SUPERSPEEDWAY
october 1998

TALLADEGA, ALA.— It seems to me that there is an annoying pessimism where Talladega Superspeedway in particular, and restrictor-plate tracks in general, are concerned.

Drivers are frustrated, but more and more of them are resigned to the idea that there is no viable alternative to the dreaded plates, which limit the horsepower and leave their cars hanging in perilous multi-car drafts. To summarize a sea of quotes, "We hate it, but the fans love it, and it's not going to change, so we might as well do the best we can and hope we don't wind up in the hospital."

Let's cut through the propaganda.

First, do the fans really love it?

Sunday's Winston 500 was played out with the usual 30-car drafts, the usual games of chess among drivers who alternated working together with deceiving each other, depending on whether they felt comfortable and whether they saw a way to advance their own cars up through the labyrinth. More than one driver got the proverbial knife in the back, some got it over and over, and if you don't believe it, cue up the video tape and focus on the Chevy of Mike Skinner and the Ford Thunderbird of Jimmy Spencer.

Dale Jarrett won this race because he is a fabulous restrictor-plate driver, a veteran who knows when to move, when to stay put and where to go. Jarrett is a fine driver generally, hardly a restrictor-plate specialist, but he would be the first to admit that the skills necessary to win at Talladega and Daytona are vastly different from those necessary to excel at other tracks.

Today we have a scenario in which 30 or more cars ride at breathtaking speed within inches of one another, and yet in 188 laps on Sunday there were only 20 lead changes. Before the latest implementation of the restrictor plates in 1988, a more common scenario would have been to have from five to 10 cars racing together at even greater speeds, with those cars swapping the lead 40, 50, even 60 times.

To this observer, the old way was better. Most of the drivers who experienced it—Dale Earnhardt, Bill Elliott, Darrell Waltrip—would heartily agree.

This resignation that nothing can be done about it disturbs me. All of the technology available today is supposed to empower, not imprison, the teams. This is NASCAR, this is America, dammit, and we're supposed to be able to do whatever we want, right?

Two factors control the competition: power and aerodynamics. The problem, reduced to a basic level, is that aerodynamics has become by far the more significant factor. The cars are too slick. They cut too small a hole in the wind; with the plates limiting the horsepower, no one has enough power to pass. As a result, a driver nowadays has to bargain with the devil, i.e., his opponents, in order to get anywhere. I know they like to call this a team sport, but gee whiz, that doesn't mean the teams are supposed to conspire with one another. At present, there is no other way to win at Talladega. All day long, they alternately help and betray each other, and the guy who manages to get help at just the right time wins the race.

It is the world's fastest, most dangerous, game of chess, and Boris Spassky doesn't even have a ride.

If it was better in the old days—and it was—then NASCAR ought to turn back the clock. This is hardly unheard of. The basic engine configurations haven't been used in passenger cars in more than a decade. One of the keys to NASCAR's popularity boom has been its relentless use of retro-technology to keep the competition intense.

So?

Make the cars less aerodynamic, which is to say make them more, not less, like the passenger cars they are supposed to reflect. Take off the front air dams and just put an old "blade spoiler" on the front, like the old days. Take a lesson from the American Speed Association, and make all the teams get their bodies from the same source. Don't allow all the wind-tunnel tinkering that subverts the use of templates. The most important factor in stability is balance between the front and rear, not slickness.

It wouldn't hurt to make the cars bigger and heavier. This wouldn't really mean that everyone would have to drive Buick Roadmasters. The production Taurus is a large car, and current NASCAR wheelbases reflect what NASCAR wants, not what Detroit produces. For instance, the wheelbases in the Winston Cup Series and the Busch Grand National division are not even the same. The Craftsman Truck Series reflects the other divisions, not the true wheelbase of pickup trucks.

Stretch them out. Bulk them up. In so doing, make them safer as an added bonus.

Narrow the tires and make them harder. Or trash the radials and bring back the bias-ply tires. Make the spoiler the same size, front and back, for all the cars. Take the darn restrictor plates off. The cars won't go any faster because they will be less aerodynamic, and maybe, just maybe, someone will be able to pass without convening a chorus line.

Will the cars be harder to drive? Yes. Will the drivers complain? Yes. Will they get used to it? Yes. Will it be fair? Yes. Will the competition improve? Yes.

Will it happen? No.

TEXAS MOTOR SPEEDWAY
april 1997

FORT WORTH, TEXAS— Buffalo Bill France brought his wild-west show to the Texas prairie Sunday, and to borrow from the cowboy vernacular, it weren't no pretty thang.

Races like the Interstate Batteries 500 have been on television before. Of course, it was a video game, the major competitors were Mario, Kooba and Yoshi, and all were driving karts.

No one has ever seen anything quite like the first Winston Cup race at Texas Motor Speedway. No one had seen anything like an atomic bomb before one landed on Hiroshima.

What is it about this track? Without question, it is one of the more attractive sports palaces on the face of the earth.

For some mysterious reason, the finest drivers in America cannot race on it. To their credit, most of them warned us. Its beauty was fleeting, they said, though not in those exact words.

Ticking off a list of what went wrong would require five chapters and 100 pages, but here is a smattering:

· Instead of a dignified national anthem by Van Cliburn, there was a disfigured one by a guy named Woody.

· The traffic situation made Napoleon's retreat from Moscow seem like a Christmas parade.

• Ernie Irvan received yet another set of stitches. Soon Raggedy Ernie and Raggedy Kim dolls will be available in Texaco stations.

• Mark "thowed a rod." Ted "blowed up." Everybody else wrecked.

• Interstate Batteries paid a gazillion simoleons to have its name on the race. A car sponsored by Exide Batteries won. In retrospect, it should have been named the Miles & Miles O' Texas.

• Texas Motor Speedway replaced North Wilkesboro on the schedule. The average speed would have set a North Wilkesboro record.

• A motor home burned in the infield.

• Tom Landry spoke to the crowd beforehand and said he hoped it would be "a good game."

Rusty Wallace climbed out of his battered Ford, surveyed the wreckage and summed it all up.

"I think they will reconstruct it and rebuild it," he said. "There's no way in the world we will come back here with the track in its current condition. I'll tell you that."

A group of fans drove from Charlotte to Texas in a dilapidated bus, suffering three breakdowns and needing four days to arrive. The bus was painted black, with a "3" on one side and a "2" on the other. If they had entered the bus in the race, it would have finished fifteenth.

WATKINS GLEN INTERNATIONAL
august 1997

WATKINS GLEN, N.Y.— One word that could be used to describe Watkins Glen International is quaint. Another is antiquated.

By the standards of modern racing facilities, the Glen is hopelessly outdated. The grandstands hold no more than 20,000 fans or so. On one side of the start-finish line is a pit area. On the other is a line of trees, under which fans line up behind a chain-link fence to watch the race.

Motor Racing Network broadcasts from a ramshackle tower that would be a pox on the face of most high-school stadiums. The entire racing surface is enclosed by light-blue guard rails. Guard rails went out of style at most circuits decades ago.

Almost everyone who attends races at the Glen camps out, either inside the track or on the adjoining acres. Bundles of firewood can be purchased at roadside stands near the track at a going rate of $5 an armful.

The huge number of campers and the small number of grandstand seats make the crowd that attends races here unique. A seat is only a small convenience, since there is no spot from which a fan can see the entire track. So they trundle through the two tunnels, dragging their trailers and pitching their tents as near as possible to a favorite turn or vantage point.

These hills are alive with the sound of unmuffled engines, roaring, sputtering and backfiring as their reverberations echo across the land.

While the Earnhardt fans are still very much in evidence, this crowd is a bit more sedate than the average stock car gathering. Because most of the fans camp out, the campers are more knowledgeable than the relentless party animals who frequent the infields of many ovals. The beer still flows freely, but one is as likely to encounter a campfire surrounded by guitar pickers as a rusted bus loaded down with roving ne'er-do-wells. The typical music here is the Grateful Dead, not Hank Williams Jr., and Budweiser might get a run for its money from the microbrews. The nearby wineries offer other options for those so inclined.

A hockey player is going to be the grand marshal for Sunday's Bud at the Glen. This is one of three tracks—New Hampshire and Michigan are the others—where both the American and Canadian national anthems are sung.

But just because this track is in the State of New York, do not confuse it with the city. The Finger Lakes area is about as cosmopolitan as a county fair. The stores that adorn the winding country roads are the same as those in Tennessee. Only the accents have been changed to protect the innocent.

This may be the most remote, sparsely populated resort area in the country. Only about 6,000 people live year-round in the county, Schuyler, which includes the race course. The lay of the land is breathtakingly beautiful, but the winters are harsh and the lure of cities like Syracuse and Buffalo is great to the young growing up here.

Broadway is hours and hours away. The only really famous thing that ever happens here anymore is the Bud at the Glen. So get your chores done, kids. Load up the (station) wagon. There's a heck of a party going on up on the ridge.

A Mark Martin fan at Dover Downs, September 1999.
Tyson Cartwright

chapter 3

FANS

THEY EVEN INVADE THE BALLPARKS
may 1998

ANAHEIM, CALIF.— Don't think the success of today's California 500 is entirely due to the Golden State's warm embrace of NASCAR. This event, like every other on the Winston Cup Series, draws fans from everywhere.

At Edison Park, the Anaheim Angels' gorgeous new home, I sat next to a couple of race rans who, like me, had decided to take in a ballgame on the side. How did I know they were race fans? C'mon. A soldier is no more recognizable than the average NASCAR partisan. The man wore a weathered Earnhardt cap, constantly updated with souvenir pins. He had a "seven-time champion" T-shirt. His wife wore a black jacket with a red Chevy bowtie logo. They were from Nashville, Tennessee. They had gone to Texas in 1996 ("It took four and a half hours to find my damned car"), been to Bristol and Charlotte innumerable times, and now they were off to the Golden State for another NASCAR-fueled junket.

By and large, they were unimpressed.

"We been out here three days, and ain't hardly heard noth-ing about the race," the man said.

"I reckon there's a lot of competition for your leisure time," I replied. "You've got to remember, this area lost two pro foot-ball teams, and it's like the folks out here didn't even notice."

"I reckon. It still don't seem right."

They were like most racing fans. The only way a newspaper could have enough information about NASCAR to please them would be if the paper didn't have anything else. To test my theory, I pointed out that the Nashville racing writer, Larry Woody, was a friend.

"He don't do a whole lot with it," was the man's reply.

"It's not him. A writer can't put something in the paper that his boss doesn't have room for."

"All I know is what I see." End of that conversation.

I asked him if he had attended the Busch Grand National race in Nashville, though, and his eyes lit up.

"Aw, yeah. We had a big time. We cooked rabbit, deer, had deer-ka-bobs, hell, we probably brought more food than they sold at the concession stands."

The man told me he was friends with Coo Coo Marlin, "Sterling's pappy," and that he remembered when Darrell Waltrip was racing go-karts.

He had not been at the speedway that day, so he asked me about qualifying. When I told him Jeff Gordon had won the pole, he frowned.

"How 'bout my man?" he asked, tipping his Earnhardt cap.

"Not in the top 25. Somewhere around 27th, I think."

"Damn."

The man had another attribute of the racing fan: a desire for close competition. He and his wife were pulling for the homestanding Angels, but when they led 7–1 and it got to the bottom of the eighth, he changed his tune.

"Get you a hold o' one, Big Frank!" he growled when White Sox slugger Frank Thomas strode to the plate.

Thomas flied to center.

"Pick ye out one ye can hit!" he yelled when Albert Belle stepped up.

Belle flied to left.

"Let's get outta here, Irene," he said. "Maybe we can beat the crowd."

THEY LOVE THEIR SPONSORS
november 1998

SOUTHERN PINES, N.C.— Things just occur to me. I don't know why.

Friday morning I was eating breakfast in Shoney's, and I noticed the guy at the next table. By observing him, I found a new perspective on the hold that racing exerts on its fans.

He was a heavyset, bearded man, probably a factory worker, truck driver or farmer. I could see him driving a pickup with a "Guns don't kill; people do" bumper sticker.

His cap and shirt both carried the logo of his favorite NASCAR driver's sponsor, and that's what drew my attention.

The collectibles made no reference to Ricky Rudd, undoubtedly the man's favorite driver. It had only the logo of Tide laundry detergent and the script letters, "Tide Racing."

Tide Racing. Laundry detergent. Could anything but a race-car driver coax this man into helping a company sell suds?

I wonder. Does he sit in the stands during a race and scream, "Go Tide! Gosh knows, I love that darn detergent!"

Does he dispatch his petite wife to the grocery store with the stern warning, "Honey, don't you come back here with All-Temperature Cheer, you hear me? No store brands, durn it! I want Tide! It whitens whites, brightens brights and don't you forget it!"

I could utilize a lot of adjectives to describe race fans, but "brand-loyal" would not be the first one to come to mind.

NASCAR is constantly citing statistics on the brand loyalty of fans, but I don't trust such statistics. They also claim that the average race fan is an affluent female attorney with political ambitions, but I haven't seen Hillary Clinton swilling Budweisers on top of a primer-painted school bus in the Darlington infield yet.

But NASCAR fans are brand-loyal. I know this not from statistics but from observation.

Some sponsors are naturals, like Budweiser, Pennzoil and Winston, because if you walk through the stands, it is obvious that the fans are beer-drinking, oil-burning, cigarette-smoking folks. RJR claims that it does not aspire to hook newcomers to smoking but that its racing sponsorships encourage people "who choose to smoke" to use its brands. Insofar as racing is concerned, this is probably true. I think more than one Jimmy Spencer fan has switched first to Camels and then to Winston, even if he hated the taste at first, just because Mr. Excitement, who does not smoke, had his paint job changed.

Once I read that Marlboro, another cigarette brand, was originally smoked mainly by women. By adopting the western motif of the "Marlboro man," and the soundtrack of the movie *The Magnificent Seven*, the cigarette's image was completely transformed. All of a sudden, Marlboro became the brand of the macho smoker.

Now, what could racing do for products with wimpy images?

Would factory workers start having "Count Chocula" each morning for breakfast, eschewing the bacon and eggs? Would kegs of Zima suddenly start appearing at a Fourth of July pig pickin'? Would *Ally McBeal* suddenly be the favorite TV show of men and women who punch clocks?

With the final pre-election weekend beckoning, if I had been a strategist for the Democrats, I would have bought me the rear quarter-panel on a Monte Carlo.

WAITING FOR THE
SECOND COMING OF DALE
october 1998

CONCORD, N.C.— Stock car racing fans are probably stereotyped more often than any humans on earth. They are often dismissed as beer-drinking, Winston-smoking, cheeseburger-eating, Republican-voting good old boys by people who have never been within a mile of a NASCAR race.

Fans will surprise you if you actually have the audacity to wade into their midst.

For one thing, there are a lot of girls amidst the good old boys. Based on an informal survey conducted this week at Charlotte Motor Speedway, the women fans are often more opinionated than the men. Women make up the most passionate supporters of Winston Cup point leader Jeff Gordon, and they are also his most unruly detractors.

For instance, Tara Zeger, a Gordon fan from Chambersburg, Pennsylvania, stood up for her hero, saying, "Jeff doesn't deserve the meanness he has to go through. I get tired of hearing people say he hasn't paid his dues. He's been working on his driving skills since he was five or six years old."

Another female fan, Regina Mull of Hughesville, Pennsylvania, piped up just as loudly with, "I hate him (Gordon). I think he's cocky. He hasn't paid his dues.

"Besides, he's a big sis."

Mrs. Mull's husband, Edward, shrugged his shoulders. "She's pretty well got it covered," he said, chuckling.

For most of NASCAR's history, Chevy fans have hated Ford drivers, and vice-versa, but Gordon's success has broken the General Motors ranks.

On Saturday morning, a tent was set up by General Motors and the United Auto Workers in which fans could climb into cars decorated to look like those of Gordon and Dale Earnhardt and have photographs taken. Both lines were long, and it was like competing armies signing up recruits. In one line were Earnhardt fans who hated Gordon. In the other were Gordon fans who hated Earnhardt.

"I can take him (Gordon) or leave him," said Steve Hamilton, an Earnhardt fan from Columbus, Ohio. "It's just from the way he wins too much. I don't see how anybody can be that good. I'd like to see somebody go toe-to-toe with Gordon. The other guys can't even run with him."

"He's all right," said Henry Young, of Petersburg, Virginia. "He's just on a roll right now. . . . A pretty long one."

"I don't like him," added Dave Wannamaker, from Middletown, Ohio. "I guess I'm a pretty hard-core Earnhardt fan, but I think Dale can still take that boy. Richard Childress Racing is going downhill since they started having these crew-chief shuffles. I don't wish [Gordon] bad luck or nothing. I just want him to get his butt beat by the 3 car (Earnhardt)."

Surprisingly, the Ford fans seemed more charitable in their opinions of Gordon. Rom Roper of Roanoke, Virginia, dubbed himself a Mark Martin fan but was quick to point out, "Gordon's all right. I'm not necessarily anti-Gordon. It's just that he's never had no bad times. I think it was Earnhardt who told [Gordon] he'd better ride that horse while he could because he's gonna break his leg on that thing before long.

"But you've got to give that kid (Gordon) credit. He's a hell of a driver whether you like him or not. When he was a teenager, I watched him driving sprint cars on ESPN, and one night I saw him take one of them sprint cars up on two wheels and then go on to win the race. You could tell even then that Gordon could flat-out drive a race car."

"You've got some people who will always hate him," said Jared Stowers, a Gordon fan from Sanford. "I think it's kind of like what happened to (Darrell) Waltrip. Most of the fans will eventually get used to him."

The last to get used to Gordon will be the Earnhardt faithful, many of whom have become resigned to the fact that their aging hero is not himself up to the task. Outside the track, business was twice as brisk at Dale Earnhardt Jr.'s souvenir trailer as it was at the seven-time champion's.

Dale Jr. has, in fact, become the fans' great white hope. The *Great White Hope* was a Broadway play, later adapted to the screen, referring to the great howl, in the early years of this century, for someone to beat Jack Johnson, the first black heavyweight champion. Many fans of Dale Earnhardt are marking off the days on their calendars until Dale Jr., the leading star in the Busch Series, arrives in Winston Cup to slay what they perceive as the Gordon dragon.

For all the animosity, however, that is reflected every time Gordon shows his freshly scrubbed face to the public, evidence exists that even the Gordon-haters are starting to moderate their stances.

"I know people who work at Hendrick Motorsports," said a Charlotte fan who works at the Richard Petty Driving Experience. "They say Gordon's a nice kid, but Ray Evernham (Gordon's crew chief) is a [jerk]."

"Gordon is a good driver, but his ego has grown a little too much as far as the way he treats the fans," said Bill Wilson of Baltimore. "He doesn't seem open and accessible.

"Yet, at the same time, I can't see booing anybody. That's poor sportsmanship. These 'Anybody But Earnhardt' and 'Anybody But Gordon' types are full of it. I think it's bad for the sport."

"Gordon's done a lot of winning," said Jeff Shuck, of Marietta, Ohio. "It's just like it always is: people turn against him."

THEY ARE A HARDY LOT
march 1998

HAMPTON, GA.— A bearded man walks down a muddy path, accompanied only by a little girl. The child is wearing only a black T-shirt and jeans despite a persistent rain. Eventually they arrive at a camp site, where a tent has been pitched on the edge of a pond. The man breaks off a piece of beef jerky and hands it to the kid. They sit on a log together, pants soaked to the ankles in mud, wretched but proud.

A Civil War battleground? The last survivors of nuclear holocaust? Bosnian refugees? Salvadorean revolutionaries evading capture in the rain forest?

Don't be silly. The man and child represent the most rugged, fierce, stubborn, feisty sub-species on earth: the NASCAR fan. Impossible to eradicate or even to price out of the market, they arrive at tracks week after week, and no action, be it economic, political or psychological, whether human or divine, will discourage them.

I have followed this sport since I was a small boy, and I love it. But enough to live in a tent through four nights of thunder, lightning and flooding? Enough to pay $35 for a T-shirt of my favorite driver? Enough to pay $150 for a decent ticket?

No. I do not love it that much.

So how, on nearly every weekend, can there be 100,000 out there who do?

On Saturday, after the day's race had been rained out, I was caught in one of those mothers-of-all-traffic-jams that frequently surround Atlanta Motor Speedway. Locked in virtual gridlock, I could observe the wretched conditions experienced by thousands of fans camped out on the grounds. Flooded campsites. Foot-deep mud. Old school buses, motor homes and minivans hopelessly snared in the red-clay ooze.

I made eye contact with one of the pilgrims tramping along the side of the roadway. I learned out the car window and asked him how he managed to put up with all this.

Somehow he managed to smile. "Beer, Skoal and Slim Jims," was the reply. Certainly not soap and water, I thought.

For all the concern about rising prices, Speedway Motorsports CEO Bruton Smith does have a history of holding back some small corner of his facilities to attract fans who otherwise might not be able to afford tickets. When it became clear that this race would not sell out, track president Ed Clark opened a portion of the back-stretch grandstands and put the tickets on sale for $20 each. In nearby Griffin, a man spent an entire Saturday riding around town, asking people if he could haul off scrap metal—rusted swing sets, hubcaps, old lawn furniture—from their yards. Eventually he collected enough junk to get $20 from a junkyard owner to buy a ticket. You'd better believe that guy figured out a way to come back to the speedway Monday morning.

These fans are indomitable. Truly they are the sons of the pioneers, glorious anachronisms in an age of remote controls, cellular phones and laptop computers. Long after the rest of civilization has given way to the cockroaches, the race fans will continue to thrive.

I am in awe.

chapter 4

OPINIONS

WHAT DO THE DRIVERS KNOW?
march 1998

LAS VEGAS, NEV.— The drivers all love Las Vegas Motor Speedway. Its racing surface is wide, gently banked and adhesive.

The drivers all love Las Vegas. Most are gamblers by nature. Nothing puts the cap on a sweaty day at the track like a few hours of recreation at the gaming tables. It's hard to lose as much money gambling as they're bound to win racing around and around the 1.5-mile tri-oval.

The fans love the speedway and the city. On the West Coast, they don't know what a real stock car track is.

By the manly standards of a Darlington, a Charlotte or a Rockingham, the Las Vegas 400 was a bunch of cars going around and around all day. It was a perfect example of what stock car racing's detractors always say about the sport.

Boring.

Like California and Michigan, its two-mile cousins, Las Vegas Motor Speedway was designed as a compromise to suit the needs of both stock and Indy cars. Like the much-maligned baseball/football stadiums of the 1960s like Atlanta-Fulton County, Riverfront, Three Rivers, this track fulfills neither purpose well.

With 20 laps to go on Sunday, a steady stream of fans headed for the exits. When was the last time that happened at a Winston Cup race?

The combination of this antiseptic environment and a 1-2-3-4-5-6-7 Ford sweep turned this inaugural into a snoozer.

Say what you will about Bruton Smith's new track in Texas or his controversial reconfiguration of Atlanta. Both tracks produce action. And it takes a superlative driving performance to win there.

Is this the future of America's fastest growing sport?

Five of the top 10 cars (nine of which were Ford Tauruses) were owned by Jack Roush. Dale Earnhardt drove a faultless race and finished eighth in the first non-Ford. The next Monte Carlo was driven by Terry Labonte; he was 15th.

Allegedly, a sellout crowd of 107,000 fans attended this abomination. Actually, there was a smattering of empty seats throughout the third and fourth turns. Every casino in town was giving away tickets. Perhaps a few of the lucky winners elected to stay home and watch on TV.

Predictably, at the end, the unabashed promoters of Motor Racing Network spoke of the fans rising to wave their arms and pump their fists as Mark Martin crossed the finish line. It was a lie. Through binoculars, peering across the track from the infield press box, most of the fans remained seated, listless and drained from several hours of noise, exposure to the sun and boredom.

Martin called it "the ultimate race track."

Then again, he also told Wayne Newton that he loved him. Let's not take either remark at face value.

BRUTON AND NASCAR
august 1997

BROOKLYN, MICH.— NASCAR is making a mistake by not allowing Bruton Smith to have a second Winston Cup date for his track in Texas.

Room has been made for Las Vegas, hiking the 1998 schedule to 33 dates. What's one more? Depriving Smith and his Speedway Motorsports empire of a race they have sought desperately for months just increases the distrust and bad feelings between NASCAR's two great empires.

No matter how often Bill France Jr. and Smith issue public disavowals ("Bill France has been a great friend," said Smith), there is a rivalry. It is a cold war that International Speedway Corporation (ISC), the France family's entity, seems to be winning at present. But there is no further need to play hardball.

The dominoes began falling first in Smith's corner. Speedway Motorsports Inc. began gobbling up tracks at an astonishing pace: parts of Rockingham and North Wilkesboro, Bristol, Sears Point, and the glittering new edifice in Texas.

If Smith wanted Daytona to take notice, he certainly achieved his goal. The slumbering giant came to life.

"Oh, so you want to play mergers and acquisitions, eh?"

ISC purchased an interest in Penske Motorsports, which has, at least apparently, added North Carolina Motor Speedway in Rockingham to its roster. ISC and Penske joined together to purchase controlling interest in tracks in Homestead, Florida, and Madison, Illinois. ISC bought the mile track in Phoenix, Arizona, and apparently will soon build a superspeedway in Kansas City, Kansas. ISC's ally Bob Bahre has signed a management contract with Las Vegas Motor Speedway. All of these properties were also sought by Smith's Speedway Motorsports.

Sources in Daytona have indicated that the France family's buying spree is not over yet.

So, if you're keeping score (and who isn't?), Smith now controls Charlotte, Atlanta, Texas, Bristol and Sears Point. Because of his willingness to schedule Indy Racing League events at three of his tracks, Smith has something of an alliance with Tony George and Indianapolis Motor Speedway.

France now controls Daytona, Talladega, Darlington, Watkins Glen, Phoenix, Homestead, Martinsville (the France family personally owns considerable stock), St. Louis (Madison is only a few miles away) and Kansas City. The alliances with Penske and Bahre give ISC considerable additional clout at California, Michigan and Pennsylvania. Most of the independently-owned tracks currently on the schedule have deep-seated loyalty to the Frances.

Suppose Smith wanted to run his own series, which he denies. NASCAR could quickly replace his spots with dates at new tracks. In any power struggle, Smith would lose, and he knows it.

In this nationwide game of chess, the France forces have the Smith forces firmly in check. Why add insult to injury? The good of the sport should now be the overriding interest.

Give Bruton his Texas date. It makes good business sense. Smith is offering to pay a $5 million purse for a second race. That ought to quiet any complaints from teams concerned about the travel expenses associated with a larger schedule. Smith could placate stockholders whose investments were predicated on two Texas races.

Do Bill, Jim and Brian France want to coexist peacefully with Bruton Smith? Or do they want to run him out of business? The release of the 1998 schedule should go a long way toward answering those questions.

THE ALTERNATIVE IS TO BELIEVE THEY'RE BETTER

may 1997

CONCORD, N.C.— In racing, as in life, no one wants to take responsibility for his own actions anymore.

Let's just suppose that Jeff Gordon wins almost every race. Do the rest of the competitors work twice as hard in a feverish attempt to overcome Gordon's advantage?

Bwahahahaha. Of course not.

No one believes it is possible that another team could be better by following the same rules it does. Hence, if Gordon is winning, he must be cheating. Either that, or NASCAR wants him to win.

Isn't it possible that Gordon is just one righteous racer? That he and crew chief Ray Evernham have forged a combination that is tough to beat?

Darrell Waltrip has seen this before. During his prime, while driving for Junior Johnson, Waltrip seldom got the credit for his victories. Junior had to be cheating. He and NASCAR were in cahoots.

"That was the most legal car I ever drove," Waltrip said recently. Even as he said it, old-timers in the back of the room began to snicker.

Such is life.

In the nauseating political atmosphere that has come to dominate major league stock car racing, each side believes what it is saying to be true. Ford engineers genuinely believe Gordon's Chevrolet, which they say has never been aerodynamically tested by NASCAR, to be superior to any other Chevy team. Chevy teams truly believe Ford has an edge in engine horsepower. Pontiac teams believe everyone is out to get them. In general, the embattled Pontiac partisans may be the most paranoid of all. But as long as they tag along behind the Fords and Chevys, seldom winning, no one listens to their strident pleas.

What is the alternative to whining? Admitting the other side is better? Never!

But NASCAR is a responsive governing body. Under Gary Nelson's direction, it is positively touchy-feely. When teams whine, NASCAR listens. Gary Nelson and his rules makers represent one gigantic lump of clay; constantly changing shape and easily molded.

On Wednesday night, Gordon suggested that NASCAR had told him he could not race the experimental chassis that carried him to victory in the Winston. NASCAR denied everything.

On Thursday, Gordon's Hendrick Motorsports associates Ray Evernham and Jimmy Johnson denied everything.

Ford's Dale Jarrett and Mark Martin called for an inspection of Gordon's car.

"Until NASCAR takes Jeff Gordon's car to the wind tunnel, I'm not even going to discuss this stuff anymore," said Jarrett.

NASCAR and Chevrolet denied everything.

Pontiac took some of Jarrett's comments as a personal insult. How dare he insinuate that they ought to be winning races! Pontiac issued a press release, accusing Ford of treachery and NASCAR of dishonesty.

Ford and NASCAR denied everything.

Remember when drivers just shut up and let their race cars do the talking?

Bwahahahahaha.

FALL: WHEN THE BIG BOYS FREEZE UP
september 1998

MARTINSVILLE, VA.— Every year the end of summer finds me longing for new stories to write, and every year the autumn provides welcome relief.

Let me elaborate. During the summer the Winston Cup Series emulates the words of Claude Rains in *Casablanca*: "Round up the usual suspects." Then, in the fall, the catch phrase becomes the tourism slogan of South Carolina: "Smiling faces, beautiful places."

During this year's summer season, from June 14 through September 20, every race but one was won by Jeff Gordon, Mark Martin or Jeff Burton. The lone exception was Jeremy Mayfield's triumph at Pocono on June 21. In 1997, from June 1 through August 31, every race but one was won by Ricky Rudd, Jeff Gordon, Mark Martin or Jeff Burton. The exception was John Andretti's Daytona victory on July 5. In 1996, Geoff Bodine's Watkins Glen victory was the only such occurrence between the beginning of the season and October 20. In 1995, Kyle Petty's June 4 breakthrough at Dover was the only anomaly until October 22.

In each case, the same old story was getting old until . . .

Ward Burton and Rudd scored late 1995 victories. Rudd, Bobby Hamilton and Bobby Labonte won for the first time in the last three outings of 1996. Hamilton scored again, along with Labonte, in the final three races of 1997. Of course, Rudd scored a long-awaited victory on Sunday at Martinsville, breaking another mind-numbing streak of "same old stories."

Fact is, these late-season breakthroughs are saving NAS-CAR from some rather embarrassing statistics. Until the leaves began to turn in 1995, we were looking at seven different winners, eight in 1996, nine in both 1997 and 1998. These would have been near-record lows for the modern era of the supposedly most competitive racing series in the world.

What explains this recurring phenomenon? A cynic would say NASCAR intervenes by giving some previously obscure performer "the call to rescue its statistics" which, of course, is nonsense.

A more likely explanation is as natural as the connection between beer and fans. The race for the championship causes a largely unconscious tendency for the leaders to become cautious. They start eyeing each other, and it gives other drivers a chance to slip into victory lane.

Why has Gordon never won a race in October or November? Well, for the last three campaigns, Gordon has been in the hunt for the title. The specter of the championship and all it entails—big money, television exposure, new endorsements, souvenir and collectible sales—becomes more important than winning races.

Winston even recognized this when it started its Leader Bonus program for rewarding the point leader when he wins a race. The sponsor wanted to make it more worth the leader's while to go for the checkered flag.

So far the bonus program has not cured the "problem," which is fine. Lone victories, late in the year, provide uniqueness and excitement that otherwise might not happen.

Who was unhappy when Rudd kept his incredible 16-year streak alive? Who would begrudge long-suffering Bill Davis and Ward Burton their only victory to date? How exciting was it to see car owner Richard Petty return to victory lane with

Bobby Hamilton? Wasn't it fun to see Bobby Labonte pull out victories on the final weekend of both 1996 and 1997?

Of course it was.

The most common question asked me by fans is, "Who do you pull for?" The answer is strictly selfish. I'm a writer, and I pull for a good story. My best friend could win three straight races, and I'd be praying for some relief in the waning laps of the next one. What? My best friend? Again? Forgeddaboutit.

Rudd's victory in the heat of Martinsville? Bring it on. As Bud Moore once screamed at me, "By God, now you got a story to write!"

ARE THE RACES FIXED?
march 1998

If a person is inclined to believe in conspiracies, he (or she) is going to suspect manipulation everywhere. If a person is cynical by nature, he is not going to believe a story that, on the surface, is inspirational.

Some people believe professional wrestling is no more rigged than any other major sport. Those people are having a field day with the NASCAR Winston Cup Series right now.

For the record, I do not believe that racing is fixed. I think there is some corruption, but I do not consider it institutional. On occasion, I think it is possible that a friendly NASCAR inspector might let one car slip through the line and stop another. I also think that, on occasion, a basketball coach like Bob Knight receives preferential treatment from an official because he intimidates him, and on other occasions, an official is biased against Knight because a Knight tantrum affects him in the reverse way. Sports are administered by humans, and humans have weaknesses.

I also think that NASCAR sometimes blurs the line between entertainment and competition in its never-ending quest to keep the competition close.

Some critics believe that the outcome of Winston Cup events has been manipulated. Namely, the suggestion has been made that Dale Earnhardt's victory on the very day of NASCAR's 50th anniversary was too perfect. The same has been alleged about Richard Petty's 200th victory, on July 4, 1984, in front of President Ronald Reagan. To a lesser degree, cynics have cited Bobby Labonte's Atlanta victory in 1996 on the day his brother Terry clinched the Winston Cup championship, and Jeff Gordon's win in the inaugural Brickyard 400 in 1994, when a triumph by a native Hoosier seemed oh-so-fitting. There are other examples, as well.

I can't buy the suggestion that the fix was in or that these victories were the result of receiving "the call" from NAS-CAR.

First, it's not like Earnhardt doesn't know his way around Daytona International Speedway. Even in 1997, when the seven-time champion failed to win a race, he could have won all four of the races on the restrictor-plate tracks, Daytona and Talladega. The man was overdue. As for the historic Petty victory, if there was a fix, no one told Cale Yarborough, whom Petty beat to the line by several feet in a bruising, screeching exchange of sheet metal. As far as the Labontes and Gordon, let's face it. No matter how fast their cars, they still had to drive them for 500 miles and finish out front. Bobby Labonte has now won three of the past four Atlanta races, and Gordon's competence behind the wheel of a race car on any track is fairly well documented.

If you believe that NASCAR's Cinderella stories are choreographed, then what about the horrible twists of fate that also occur? Did NASCAR want Bobby Allison's family to fall apart? Did NASCAR want Alan Kulwicki and Davey Allison to die tragically? The series of events leading up to Neil Bonnett's death were lifted right out of a Greek tragedy, but has anyone suggested that NASCAR was out to get Bonnett? Did the France family want Tim Richmond to contract AIDS or Rob Moroso to die in a traffic accident? Of course not. Amazing stories happen regularly in this sport, both good and bad.

The above having been duly noted, I do understand why certain fans are suspicious. The qualifying results last week at

Atlanta were, at the very least, a bit strange. When one starts adding on layer after layer of coincidence, they get stranger.

Here are the basic facts. Todd Bodine, who had failed to make the starting field in the season's first three races, qualified second. The third-place qualifier, Dick Trickle, had started 34th, 15th and 12th in the season's first three races. Robert Pressley, who started seventh, had qualified 30th, 26th and 20th.

Rumors were circulating everywhere that Bodine's new sponsor, Tabasco, was growing rapidly displeased with its lack of exposure. NASCAR likes to keep its sponsors happy. Trickle's car owner, a wonderful fellow by the name of Junie Donlavey, spent the weekend in a Richmond hospital undergoing open-heart surgery. Surely Trickle's qualifying performance boosted Junie's spirits, and as NASCAR tells us constantly, the sport is one big family.

OK, phase two. Neither Bodine nor Trickle had shown in the day's brief practice session any indication of being a contender for the pole. Both were, in fact, well off the pace and seemed in danger of not even making the field, let alone qualifying in the top 10.

Phase three. Both drivers explained their sudden improvement by saying that, during practice, something had been wrong with the carburetors on their engines. It was like someone said, "But how are we going to explain this?" and the answer was, "Uh, hmm, I don't know, tell them the carburetor was fouled up."

Look, I see coincidences every day. Just Friday I was watching the NCAA basketball tournament, and a player for Valparaiso named Brice Drew missed an easy shot that could have cost his team a first-round game. Seven seconds later, Valparaiso executed an incredible length-of-the-court pass, a player lateraled the ball to Drew in mid-air, and, you guessed it, the same guy, who had been a goat seconds earlier, nailed a three-pointer to upset Mississippi. And, oh yeah, Brice Drew's father, Homer Drew, was the Valparaiso head coach.

Just as sure as I'm sitting here, there were fans watching

that game who said, "Hey, man, no way that could happen. It had to be fixed."

Human beings do incredible things, particularly when under pressure. Some come through in the clutch; others choke. In racing, some engines choke; clutches fail. Something like that.

HYPOCRISY MADE THIS SPORT WHAT IT IS TODAY
september 1998

MARTINSVILLE, VA.— Winston Cup stock car racing is about like everything else these days, which is to say it is rife with hypocrisy.

Frankly, the soap opera being acted out daily in Our Nation's Capital is no more tacky than the doings in America's Fastest Growing Sport. Being America's Fastest Growing Sport carries with it certain unreasonable demands, and the rough-and-ready fighter jocks of NASCAR have way too many twinkles in their eyes to live up to the demands being made on their collective morality.

Race-car drivers are like golf pros. And traveling salesmen. And rodeo cowboys. A lot of them have wandering eyes. Any walking tour of the garage area reveals women in ample quantities for whom wandering might be considered unavoidable.

That same garage area has few secrets. A few years ago one driver showed up at North Wilkesboro with a shiner, which he chalked up to a softball injury. Of course, the Mrs. wasn't at the track, a moving van was parked in front of the house, and the driver in question hadn't played softball since all he could do was throw underhand.

Then there was the driver who attended a glittering, black-tie-only function accompanied by "a woman who was not his wife." A variation of Murphy's Law often applies to such affairs. When one appears in a visible place with one's mistress, one's mistress will get bloody drunk and start loudly questioning the pedigree and manhood of very important persons. Predictably, the word came down shortly afterward from the Exalted Czar in Daytona for the driver to "lose the tart." When the tart was lost, all sorts of complications arose, some requiring the assistance of barristers to resolve.

Some of these stories may not be true, but they sure do get wide circulation.

Of course, this is a family sport, as we are constantly being reminded. Unless we switch the channels every time the announcers go to a commercial, this family-sport motif will be thrust upon us. All sorts of visual images flood our psyches, telling us about these folk heroes, their deep Christian lifestyles and their single-minded devotion to the fans. According to the propaganda, each week thousands of small kids have their caps autographed in the shadow of the race rigs.

I've got news for you. They don't even allow small kids in the garage. And some of the folk heroes are more likely to snarl at the fans than sign their autographs.

Before you self-righteously decry this kind of behavior, remember the high standards against which these men are being measured.

They didn't sign up to be missionaries or Boy Scouts. Most of them grew up tough and mean. They cut their teeth on dirt tracks by ramming heavy metal fortresses against other similarly equipped vehicles. Some of them have had their teeth straightened and their dialects sanitized and been lectured on the finer points of good public relations.

Thank the good Lord for a safe race. Say "we" instead of "I." (Example: "We hit the wall.") Refer to oneself in the third person. (Example: "Ricky Ruffbutt has to do what's best for Ricky Ruffbutt," said Ricky Ruffbutt.)

Want to know the high-dollar advice of most public-relations firms? One word version: Lie. Five-word version: Lie like a sewer rat. If that's not a recipe for hypocrisy, I don't know what is.

MESSAGE FROM NASCAR:
TWO LEAGUES WON'T WORK

february 1999

DAYTONA BEACH, FLA.— The future of NASCAR is shrouded in secrecy, maybe even mystery, but one thing the members of the France dynasty wants you to know is that there will not, repeat not, be a split of the Winston Cup Series.

Their friendly rival, Bruton Smith, has pointedly called for a split of stock car racing's premier series. Smith would also like a second date for his tracks in Texas and Nevada.

"The problem with League A and League B," said NASCAR president William C. France, "is who gets to be in A and who gets to be in B. That's what Bruton doesn't consider when he starts proposing a split of the series."

NASCAR has convened "focus groups" to ascertain how the fans feel about proposals to manage the sport's growth. According to Brian France, heir apparent to his father, a split of Winston Cup is one move the fans would not abide.

"Let me tell you another big problem with the concept of two leagues," said Bill France. "How do you determine who goes where? Odd numbers and even numbers? Let's say we had Bobby Labonte in one league and the race at Texas holding a race in the other. Here's what would happen. Joe Gibbs would say he needed Labonte in the other league because his sponsor, Interstate Batteries, has a bunch of employees in Texas.

"So Gibbs would go to Mike (Helton) about it, and Mike would go along with him. Then there's another team, and then another, and before you know it, it would be impossible to maintain any kind of balance."

The elder France, whose late father William H. G. France founded the governing body, talked at length about further expansion of the series.

"The only thing you cannot replace is the driver," said Bill France Jr. "We could run 52 weekends a year. The increased revenues from those races would enable teams to hire personnel, buy equipment, do whatever it takes to run those races. You could train a second crew chief. But the driver is the problem you face."

France said a better solution than separate leagues would be having a team championship in which owners would have the option of putting an alternate driver into their cars on selected weeks.

"Some of the drivers would run all 52 races," said France. "Some teams would have one driver run 40 and the other driver run 12. Maybe some teams would split the schedule right down the middle."

"When we give the drivers a weekend off now, we've got guys like Ken Schrader who get on a plane and drive in three other races that very weekend," said Mike Helton, the vice president of competition.

For the umpteenth time, France expressed his loyalty to the existing tracks where "we've been running for 35 years." He and Helton even allowed as how the chances of the existing Gateway track near St. Louis and the planned Anne Arundel County, Maryland, track have practically no chance of ever securing a Winston Cup date.

In NASCAR's view, Baltimore is a market already glutted by nearby tracks in Richmond, Virginia, Dover, Delaware, and Long Pond, Pennsylvania. NASCAR would even protect its Busch Grand National date at Nazareth, Pennsylvania. Pressed on the topic by a Baltimore sportswriter, Helton said, "There are other kinds of racing in this country they can go to. We've tried our best to let them know we have no plans there."

The St. Louis area will be more than covered, to NASCAR's way of looking at it, by the track being built in Kansas City. Gateway, of course, will retain its Busch Grand National and truck dates.

HALL OF FAMERS? I THINK NOT
november 1998

AVONDALE, ARIZ.— I feel like the Grinch who stole Christmas. Really, I'm a compassionate guy, but the events of this week compel me to take a hard line.

The latest inductees into the International Motorsports Hall of Fame were announced. Among them were Louise

Smith, the first woman driver to be inducted, and Wendell Scott, the first African American.

Although I have never met her, Louise Smith is by all accounts a pleasant, gracious woman. As a kid, I revered Wendell Scott, getting my picture taken with him in the infield of Greenville-Pickens Speedway on several different occasions. Personal feelings of warmth to the contrary, neither has any business being in a reputable motorsports hall of fame.

Mrs. Smith never won a race in NASCAR's Strictly Stock Division, which later became known as Grand National and then Winston Cup. Scott won one race, in Jacksonville, Florida.

Some would argue that both should be inducted because they were pioneers. Here is what I would—in fact will—argue.

In order to be a pioneer, one has to pave the way for others. There were as many woman drivers in Smith's day—Sara Christian and Ethel Flock Mobley, for instance—as there are today. And of course, not even Scott's modest level of stock car racing success has been duplicated by another black driver.

Jackie Robinson did not make the Baseball Hall of Fame strictly because he was the first black man to play in the major leagues. He made it because he was a great ballplayer. Did Hank Greenberg earn enshrinement by virtue of being Jewish? No. Was Roberto Clemente a hall-of-famer because he was Puerto Rican? Of course not.

Maybe, if Smith and Scott had been given the opportunity of a Jackie Robinson, they might have had hall-of-fame careers. If Babe Ruth had lifted weights, he might have hit 1,000 home runs. But the record lists him with 714. Athletes must be measured by what they did, not by what they might have done.

I am, relentlessly, a hard liner. I do not believe Shoeless Joe Jackson should be a baseball hall of famer because, at the very least, he went along with teammates' plans to throw a World Series. Pete Rose bet on baseball. I cannot forgive him for that, despite the fact that he had more hits than anyone in history.

So how in the world am I supposed to smile and say, "Louise Smith made the International Motorsports Hall of Fame. How nice. It's about time there was a woman driver"? Or, "Wendell Scott. I saw him drive around and around the apron at Bristol for five hours one day in 1965. It's about time he made it."?

Scott ought to be remembered warmly as an exceedingly nice man who stubbornly fought the odds and earned the acceptance of the fans of his day. But, except for that one dusty race in Florida, Scott did not succeed. He spent decades fighting an uphill struggle, one he did not win. That he survived is enough for him to earn our admiration, but it is not enough to place him in the select company of Alain Prost, Harry Hyde and Gordon Johncock, the others who will be inducted along with Scott and Smith next April.

Not only that, but Smith and Scott ought not be selected over such eligibles as the great Indy-car driver Jimmy Bryan, Grand Prix and CanAm ace Denis Hulme and the late Winston Cup champion Alan Kulwicki. The ballot had dozens of names on it more worthy than Smith or Scott.

Who's next? Marty Robbins, the first country singer to drive stock cars? Actress Crystal Bernard, because she drove in a celebrity race? Paul Newman, because he won both an Academy Award and a Trans-Am race?

Come to think of it, Newman does deserve to be inducted into a motorsports hall of fame ahead of Smith and Scott.

To quote Strother Martin in one of Newman's films, "What we have here is a failure to communicate."

HEY, NASCAR, LET 'EM PLAY
september 1998

DOVER, DEL.— For all of NASCAR's power and resources, the monolithic governing body is powerless to prevent a handful of teams from dominating its flagship series, the Winston Cup.

How ironic it is that four powerful car owners—John Hendrick (11 victories), Jack Roush (8), Robert Yates (2) and Joe Gibbs (2)—have accounted for the victories in 24 of the season's 27 events. This in an age in which NASCAR micromanages the sport in the name of close competition.

The waning years of the millenium find NASCAR's leadership proclaiming obedience to the legacy of founder William H. G. France, citing "Big Bill" and his dedication to close competition. France, it is said, wanted every race to end with a Ford, a Chevy and Pontiac side-by-side, inches apart at the line.

So, at the behest of vice president Mike Helton and Winston Cup series director Gary Nelson, the rules are constantly juggled and rearranged, the better to keep those Tauruses, Monte Carlos and Grands Prix close to each other. The rule book means nothing. If they want to do something that is not prescribed there, they issue a release claiming they changed it and forgot to get the changes printed. They cite the dreaded "EIRI" (except in rare instances) clause, where the rule book effectively commits suicide.

NASCAR's own rule book stipulates that a rookie cannot serve as a relief or substitute driver. But when George Elliott succumbed last Thursday, his son Bill promptly decided not to race at Dover, opting instead to put a promising young driver named Matt Kenseth behind the wheel of his McDonald's-sponsored Ford. Kenseth, having never driven in a Winston Cup race, is not even a rookie, but Bill Elliott went to Nelson, and Nelson said O.K. A rookie cannot serve as a substitute driver "except in rare instances," which has been further amended to mean, practically, that Mr. Magoo can drive if Nelson says he can.

Bill France, the father of the present NASCAR czar, believed in close competition, but he also wanted the cars to look like the cars people drive on the highway. The original name of the Winston Cup Series was the Strictly Stock Division. Originally the differences between a street car and its NASCAR equivalent were supposed to be necessary evils enacted in the name of safety.

Yet today a race-ready Ford Taurus looks no more like its street equivalent than a Contour looks like an Escort. A similar analogy can be applied to a Grand Prix and a Monte Carlo. These are tube-framed, rear-wheel-drive, doorless race cars, similar to street cars only in shape, and increasingly dissimilar even there. Their engines are like nothing used in a street car for the past 15 years.

Yet for all this blithe micromanagement, the control freaks of Daytona Beach cannot succeed. Thirty years ago, when a model year changed and one car arrived at Daytona with an aerodynamic advantage over the other, the engineers for the other make worked harder on their engines. Eventually, they achieved with power what they could not surmount in wind resistance.

Nine different drivers have combined to win all the races, and there are only seven races to go. Ten years ago, without all of the hopelessly complicated templates, the chassis dynos and the spoiler heights and angles, 14 different drivers won races, among them Phil Parsons, Lake Speed and Kenny Schrader.

Except in rare instances, NASCAR ought to leave this sport alone and let 'em play.

HOW OLD IS TOO OLD?
february 1999

If NASCAR were a nice restaurant, and a career were a waitress standing over an entree with a shaker of fresh-ground pepper, most drivers would never know when to say when.

Neil Bonnett loved driving a race car. He sustained injuries in 1990 that were originally considered career-ending. More than three years later, Bonnett, one of the more popular men ever to compete in NASCAR, got a second opinion and resumed his career. On February 11, 1994, he was killed instantly in a Daytona practice crash. He was 47 years old.

When is a driver too old? There is no simple answer. Harry

Gant was closing in on 53 when he won his final race on August 16, 1992. Certainly it depends on many factors, including physical conditioning, injuries and level of activity.

One fact that seems to be irrefutable, however, is that the last person who should choose the proper time to step aside is the driver himself. He is hardly objective. A series of interviews revealed that almost every retired driver will now admit that he waited too long to retire, and that almost every still-active oldster insists he is as good as he ever was, or at the very least, to paraphrase Darrell Waltrip, good enough to win another race.

Waltrip, his wondrous career record being perpetually undermined by a losing streak that dates back to September 6, 1992, gets tired of answering questions about why he keeps soldiering along.

"Why can't people understand that I still love driving a race car?" asked Waltrip. "I know I haven't won in a while, and no, I realize that I'm probably not the driver I once was, but I still enjoy it. I think I can still win another race. It's something I've dedicated my whole life to, and the day I stop loving it is the day I'll hang up my helmet."

Forget about sentiment for a moment. Let's look at the facts surrounding the careers of some current drivers. Waltrip, of course, has won 84 times, but the last victory was a whopping 195 races ago. Incredibly, this is not even close to the longest current streak. Dave Marcis, who will turn 58 on March 1, has started 448 races since his February 21, 1982, victory at Richmond.

Dale Earnhardt, who will turn 48 on April 29, has won once in his last 93 starts. Thanks to last year's Daytona 500, Earnhardt has 71 career victories, sixth best all time. Earnhardt remains driven by a desire to set the all-time record with his eighth Winston Cup championship. But, as the late Gerald Martin wrote last year, ". . . I doubt that he (Earnhardt) is driven by the same hunger to grit his teeth and go for broke in certain situations, the way he did as a younger man."

How could he? Two years ago Earnhardt survived one of

the most frightening crashes in motorsports history at Talladega, his stricken Chevrolet run over by three other cars as it lazily wobbled across the track. Again, to quote Martin: ". . . Earnhardt has reached that inevitable point in his career when he can tell himself that the will to win is still there. But the reality is that a hungry man strives harder for a morsel of day-old bread than does a well-fed man for an after-dinner cordial. And Earnhardt is a well-fed man who, as much as he loves the heat of battle on the track, also relishes the empire he has built. . . . Earnhardt is, I guarantee you, as ornery and as competitive as ever. But time, broken bones and incomparable success take a toll on body, mind and spirit."

A year ago, when Earnhardt won his first Daytona 500 after 19 fruitless tries, it seemed as if the stars had been put back in place in the NASCAR constellation. Sentimental observers started predicting yet another championship for Earnhardt.

By the end of the season, it seemed as if the Daytona victory had occurred decades earlier. Earnhardt wound up eighth in the point standings, marking his fourth consecutive year of decline. He had only five top-five finishes. Larry McReynolds was awkwardly moved over to Richard Childress' companion team. Nothing worked.

Earnhardt himself declines to discuss such things as the trauma of growing old, at least not seriously. He still thinks he can win a championship. He will know when it's time to quit. It is a frequent, if tired, refrain.

Waltrip, on the other hand, has a soul that is quite a bit more accessible.

Describing his philosophy on running the frightening high banks of Talladega, Waltrip said, "You are in bumper-to-bumper, side-by-side, door-handle-to-door-handle, three-deep and sometimes four-deep traffic all the the time. The intensity is incredible. The bottom line is that mentally it just wears you out."

As ex-driver A. J. Foyt said, "Racing's like rodeo. You've got to jump back on the horse after it throws you."

But that doesn't apply to forever.

CART star Bobby Rahal underwent a series of winless years before he called it quits at the end of last year.

"The whole point to my retirement was because I feel I'm ready to make the move to focus solely on the business side and the sporting side of CART and the business side of our team," Rahal said. "I made a commitment to myself and my family that this would be it.

"Frankly, for me, once you've driven these cars, which are the most exciting cars in the world to drive, I'm not sure I could get motivated to drive anything else. . . . There are going to be times when I miss racing. There's no question about that. I'd be lying . . . if I didn't think there were going to be times that I missed it. But that hasn't been the criteria. The criteria for me has been my value to my team. . . . It's time for me to give a little bit back to (my family) and try to make up for the last 25 years."

Except for a mid-season replacement stint as a substitute driver for injured Steve Park—he had a fifth and a sixth—Waltrip's 1998 season was humiliating. Waltrip found himself, for most of the year, in a predicament almost unimaginable for a man of his talents and history: puttering along in bottom-rung equipment.

"I thought I was going to have to retire whether I wanted to or not," Waltrip noted, softening the bitter truth with a chuckle.

Waltrip used 20 ex-champion's provisionals, meaning that he started at the very back of the field in 61 percent of the races.

"It's what I know I can do," he insists. "Driving a car is a piece of cake. Driving the car, running up front, that's what I know how to do, when given the opportunity. And I finally have, I think, that opportunity and a chance to do that.

"I may not be able to win every race, but I know there are races I can win," he said.

Waltrip has a rather impressive new lease on life. He parlayed the substitute stint for Park into a ride in Travis Carter's new second car, and everyone wishes him well.

Forgive a morbid observation, but they all wished Neil

Bonnett well, too. Let's just hope, that amidst the seemingly perpetual careers of Waltrip, Earnhardt, Marcis, Dick Trickle and Morgan Shepherd, there is no tragedy lurking around some fateful corner.

THE CHAMP OUGHT TO BE
THE BIGGEST WINNER
july 1997

LONG POND, PA.— Jeff Gordon has grown as a person by leaps and bounds since he arrived as a rookie on the Winston Cup Series in 1993. Where once he pretty much recited the company line—said what his handlers told him to—he has become an infinitely more interesting person recently.

So why the politically correct position on the Winston Cup point standings?

In case you haven't noticed, NASCAR's No. 1 series is governed by a system that rewards consistency at the expense of excellence. Entering today's Pennsylvania 500, Gordon has won seven races—more than three times as many as anyone else—yet trails teammate Terry Labonte by three in the season standings.

Three drivers have two victories. Labonte is not one of them. He has zero victories. Zilch. Nada. He leads the season standings. He is the defending champion. Last year he won two races. Gordon won 10.

So what does Gordon have to say about all this? He says he agrees with the point system and does not want it changed.

What? Can't Gordon see that NASCAR's crazy, consistency-is-everything formula has already cost him one title and is probably going to cost him another?

Gordon says he wants his career to be judged on the same basis as all the champions of the past. Someone had to tell him to say that. Someone with one of those gaudy Winston shirts.

The Winston Cup point formula is hardly immutable. It's not like the 90 feet that separate the bases in baseball or the 10 yards it takes to get a first down in football. It's more like the distance of a three-point shot in basketball. Over the history of NASCAR, like everything else, the formula has changed repeatedly, most recently prior to the 1975 season.

Though it is unlikely ever to be applied to stock car racing, Formula One has it just about right. On the worldwide Grand Prix circuit, only the top six drivers get points. First place gets 10, second 6, third 4, fourth 3, fifth 2 and sixth 1. The formula is simple: 10–6–4–3–2–1. Simple yet perfect. To be the World Driving Champion, you've got to win races.

If that formula were applied to the NASCAR Winston Cup Series, Gordon would have 80 points. Labonte would have 33.

The NASCAR system is designed not to reward excellence but to keep the race close. No matter how many victories one driver has, no matter how thoroughly he dominates, there is going to be a mathematical race as the season winds down.

The system does not harm the quality of competition at season's midpoint. Everyone, even Labonte, is still racing to win. But when the final five races roll around, the leaders have to go into "stroker mode." Engines are detuned so they will not fail. Car owners and spotters constantly preach caution to the drivers involved in the race for the title. The almighty championship becomes more important than the race they are running.

It is unfair to criticize a driver for "stroking his way to the title." With a $1.5 million payoff at stake, it is the only way to go.

Labonte stroked his way to the title. Gordon did so during his championship season. So did Dale Earnhardt, Alan Kulwicki, Rusty Wallace, Bill Elliott and everyone else who has won the championship in recent years.

At least make a minor change. Give the winner of a race 15–25 more points than anyone else. Make it worth his while.

Would a change in the system have cost Earnhardt any of his seven championships? No. If the system had been different, Earnhardt would have driven differently. That's how great Earnhardt is. He probably would have won several more.

By all rights, Gordon should be well on the way to his third title in a row.

In short, Gordon ought to stop praising a system that is burying him.

Don't be a patsy, Jeff.

CHARLOTTE'S PLACE IS SECURE—
FOR NOW
october 1997

CONCORD, N.C.— Remember the old television commercials in which we were told that a certain chain of food stores was "as much a part of [our lives] as the ACC"?

The commercials could have been talking about stock car racing in the Greater Charlotte area, where, it seems, there is a souvenir stand on every block, a track in every town, an Earnhardt sticker in every rear window.

Not everyone in the area loves NASCAR, but NASCAR loves everyone in the area.

A trip to the supermarket reveals Mark Martin's smiling likeness on virtually every one of another chain's store brands. From chocolate-chip cookies to charcoal, there is the hardworking Arkansan, "driver of the No. 60 Winn-Dixie Ford Thunderbird."

A year ago, Jeff Gordon could not drink enough Coca-Cola. Mysteriously, an off-season change of palate occurred. Now the Kid is guzzling Pepsis from sunup to sundown. After all, they were "born in the Carolinas."

Dale Earnhardt sells his own Chevys up near Hickory, but he also developed a recent affinity for Burger King Whoppers. Terry Labonte's deep-blue eyes beckon thousands of soccer moms to the wholesome goodness of Kellogg's Corn Flakes. Darrell Waltrip peddles parts at Western Auto. Little brother Michael steers us toward Citgo when our gauges are low. Quaker State takes the credit for all three of Rick Hendrick's cars, suggesting that its motor oil was largely responsible for the 1–2–3 finish at Daytona.

Why do we pay attention to all this ultra-commercial nonsense? First, because it cascades across our psyches like an avalanche of images. Second, because we are in The One True Stock Car Racing Capital. NASCAR pumps millions into our economy, and if we have to pay an extra nickel for a box of Tide just to keep Ricky Rudd in the lead draft, it's the least we can do.

But will this love affair with that omnipotent commercial

image, NASCAR, continue perpetually? Could this spruced-up dame—she no doubt looks every bit as gorgeous as Miss Winston—start two-timing us? Is the lure of other bright lights and other big cities too tempting to the area's one true love?

Not any time soon.

NASCAR may sanction a hillclimb up Mt. Everest, but most of the Sherpa guides are still going to come from the Charlotte area, experts say.

"A lot of people have wondered, with expansion to Texas and other places, whether this would remain as the capital when it ceases being the geographic center," admitted H. A. "Humpy" Wheeler, president of Charlotte Motor Speedway. "I don't think so.

"If you went out to Los Angeles in 1963, every Indy car was built within three miles of the L.A. airport. That's gone. It moved to Indianapolis, it left there, and eventually it went overseas. That happened primarily as a consequence of technology. But we're not a high-tech race car. I don't see that changing."

The flow—of people, parts and resources—is the key to preserving the status quo. In much of the country, motorsports is a part-time vocation. Even at the highest level of West Coast stock car racing, the Winston West Series, most of the drivers and mechanics have day jobs. When they decide to make a bid for the big time, or at least the full time, they head to Charlotte. The same is true in the ultra-competitive Northeast, where the modifieds rule hearts but not pocketbooks.

From the West, in recent years, have arrived Ernie Irvan, Marc Reno, Ron Hornaday, Mike Skinner, Chad Little, Derrike Cope and many others. From the Northeast streamed Ray Evernham, the Bodines, Ricky Craven, Randy LaJoie, Steve Bird, Mike McLaughlin, Jeff Fuller and enough others to "pahk every cah" in Gaston County.

The cutting edge is here. The Craftsman Truck Series is more geographically balanced than the Winston Cup, with races in places like Monroe, Wash., Denver, Colorado, and

Flemington, New Jersey. When the series began in 1995, a number of teams were headquartered on the West Coast. Yet in three seasons to date, only one truck from anywhere other than the South has won a race, that being a Butch Miller win with a Midwest-centered group in the first season. Now all the truck teams are flowing toward Charlotte like a rain-swollen stream.

"Now you see all these transporters in the garage area, whereas 20 years ago the Wood Brothers were still bringing their car in here on a 'low boy,'" said Wheeler. "Twenty years from now, we'll be flying these teams all over the place on some super cargo plane. They'll have the cars in there and a smaller rig with all the tools. Russia's got a cargo plane right now that you could put the whole field for the Coca-Cola 600 in. You can go where you want to go. You want to go to Japan in a few hours? That's fine."

Better lengthen those runways over at Douglas, boys. The day is coming when they'll head 'em up and move 'em out.

NEVER LET THE RULES GET IN THE WAY OF THE SHOW

june 1998

RICHMOND, VA.— NASCAR has a rulebook with one supremely important rule: "Do not let rules get in the way."

In the final laps of Saturday night's Pontiac Excitement 400, the men in the big, glassed-in bubble made one decisive call after another. There was nary a historian in the bunch. They ran this race by the seat of their pants. They paid no attention to rule of law or established precedent. This was frontier justice in all its glory.

What a show.

This sport is growing so fast that it has burned its bridges.

Analyzing the end of the race might have revealed, oh, maybe six hypothetical outcomes. In five of them, Dale Jarrett would have won. But the most wildly exciting ending had Terry Labonte bumping and grinding his way past Jarrett.

That's the one they went with, baby.

The red flag is an age-old means of stopping a race for safety reasons, used when a track is blocked by disabled vehicles or covered with oil. Rain may cause red to fly from the flagman's stand. A wall may be damaged or the pavement rutted.

A couple of observers with about a century's combined years of watching stock car races had never seen a red flag waved to prevent a race from ending under caution.

But every day is a new day. The past no longer matters in the heady, precedent-shattering dream world of America's fastest growing sport. At Richmond, there were 103,000 people to send home salivating like St. Bernards.

Let's briefly recount the events that brought down the curtain on Richmond's night-time extravaganza.

Six laps to go, with Jarrett's red, white and blue Ford pulling away. A crash occurs. The likelihood, at least to observers who have actually SEEN a race before, is that the race will end under caution, with Jarrett winning.

B-O-R-I-N-G.

It took our intrepid scriptwriters at NASCAR three caution laps to come up with their latest innovation: Hey, guys, let's throw a red flag. Then we'll have three laps of spine-tingling, All-American, green-flag racing.

Jarrett pulls away again on the restart, this despite the fact that his tires are more worn than the drivers—Labonte, Ken Schrader, Rusty Wallace—pursuing him. As the leaders streak down the back stretch, another crash occurs.

OK, now there will be two laps to go, and a yellow flag, still with Jarrett leading.

B-O-R-I-N-G.

Quick, comes the cry from the big glassed-in bubble. Tell 'em to hold that yellow flag a lap.

So, despite the lingering debris of what had been Johnny Benson's yellow Taurus, the leaders go another lap, and sure enough, Labonte bops Jarrett in the fourth turn and leaves the left side of Jarrett's Taurus considerably less red, white and blue than it had been.

Now! scream the boys in the bubble. Now! Throw that yellow flag, right there along with the white one. We've got us some highlights that are going to be shown from coast to coast.

Jarrett bumped Labonte after the race was over. Wallace raced him a lap under caution. And why not? All other established rules of order had been violated. Why not give something else a try?

Here's what NASCAR could have done. In advance of this fiasco, the governing body could have held a press conference in which Gary Nelson could have said: "The fans want races to end under green, so we have decided to use the red flag as a means of doing this. That way, the race won't have to be extended, and fuel mileage will no longer be an issue."

But being above-board in its conduct of the sport is as alien to NASCAR as it is to our President's conduct of the affairs of state.

Besides, the race did NOT end under green anyway.

NASCAR could have taken it a step further. When Labonte shoved Jarrett out of the way, instead of waving yellow and white flags in tandem, they could have ordered red and white. They could have stopped the cars again and had another battle royal.

All things are possible nowadays.

ROAD RACERS FACE LONG ODDS
may 1997

TALLADEGA, ALA.— What in the wide, wide world of sports possesses guys like John Andretti and Wally Dallenbach to stick with it?

Both Andretti and Dallenbach are members of distinguished American racing families. Andretti's uncle, Mario, is one of a handful of men with a legitimate right to be considered the greatest American racer ever. His godfather, A. J. Foyt, is another. First cousin Michael is a perennial contender on the CART open-wheel circuit. Dallenbach's father, also named Wally, is an ex-Indy car driver who now serves as CART's director of competition.

John and Wally share a more dubious distinction, as well. Neither has ever won a NASCAR-sanctioned stock car race, and neither seems likely to do it this year. Both have pretty much been stars in every other form of wheel they've climbed behind. Dallenbach's forte is road racing; he has won everything from the Rolex 24 at Daytona to the SCCA (Sports Car Club of America) Trans-Am championship. John Andretti has raced—and won—in everything from Sprint cars to Indy cars. He even made a splash behind the wheel of a Top Fuel dragster a few years back.

Granted, in the domestic motorsports universe, the Winston Cup is by far the brightest star. But why would you continue to be an average Joe in a field of 42 stock cars when you could snap your fingers and be leading the pack in an IRL (Indy Racing League) or Trans-Am event?

The answer, in Andretti's case, is twofold, according to his father, Aldo.

"The first reason is that my boy is motivated by competition," said Aldo Andretti, Mario's twin brother. "He wants to be the best, and right now, this is where it's at in motorsports in this country. I've always advised John to be versatile, to go where opportunity takes him, but it's hard to make an argument right now that anyone would be better served by competing anywhere else."

Aldo Andretti, who bears a compelling likeness to his son, gave up his own driving career to work alongside Mario in a variety of roles.

"The second reason that John is determined to succeed down here is that he just loves stock car racing," said the father. "When he was 12 years old, he used to watch the Daytona 500 and say that is where he wanted to race. Even when he was successful in open-wheeled cars, he always had this urge to drive the stock cars."

A considerable body of evidence exists to suggest that making the move from other forms of motorsports to NASCAR is harder than the reverse. Steve Kinser, perhaps the most brilliant Sprint car racer ever, was humiliated during a brief stint in the Winston Cup Series in 1995. Robby Gordon, a former

CART race winner like Andretti, continues to struggle in Felix Sabates' latest gaudy experiment.

"There are only a few road racers whom I think are capable of moving over, say, from Trans-Am to NASCAR," said Buz McCall, now a Winston Cup team owner but also a long-time winner at the SCCA Trans-Am level. "I am sure Scott Pruett could do it, because he is a racing son of a gun. Tom Kendall, I think, could be successful. Ron Fellows, yes. And if Jack Baldwin had ever gotten a shot, I think he would have made an excellent stock car racer.

"But, in my opinion, there are many, many stock car drivers who could succeed in virtually any kind of race car they wanted to climb into. (Dale) Earnhardt, Rusty (Wallace), certainly Jeff Gordon, both Labontes, Bill Elliott, Mark Martin—there's no doubt in my mind that every single member of that group could win at every level imaginable," added McCall.

On the surface, riding around an oval in a stock car seems to be a less developed skill than the rapid-fire gear shifts and frenetic turns required to master a road course.

Looks, the evidence suggests, are deceiving.

NASCAR TAKES OVER BASEBALL
may 1997

CONCORD, N.C.— Charlotte Motor Speedway held a blockbuster press conference on Wednesday. In fact, it was so big they moved it to nearby Blockbuster Pavilion. It was so big Garth Brooks was the lead-in act. As Dudley Moore might say, this was not a small announcement.

NASCAR, it seems, has actually outgrown stock car racing, and we're not talking just theme parks, restaurants and stores at the mall, either.

Major league baseball acting commissioner Bud Selig came to the speedway to announce that NASCAR had signed a management contract to run what used to be known as the national pastime.

You know, back before there was an Internet.

Selig called the merger between baseball and racing "a natural."

"Management is firmly in control of this sport," said the Budster. "Every track is wildly profitable, the athletes, compared to ours, don't make beans, and the fans cheerfully pay 80 bucks a ticket.

"When our selection committee met most recently to review possible options for a new commissioner, some representatives of the Atlanta Braves started talking about stock car racing, one thing led to another and . . . well, here we are."

Selig was asked why major league baseball's owners would be willing to turn over their sport unconditionally to the France family.

"Well, NASCAR is run the way baseball used to be," said Selig. "When I look at Billy France, I see Judge Kennesaw Mountain Landis."

Landis, you'll recall, was the baseball commissioner most noted for banning eight members of the 1919 Chicago White Sox for life, this despite the fact that a jury acquitted them for throwing the World Series.

France, standing nearby, was asked to describe his first impression when Selig approached him about running baseball.

"I told Bud, 'Me and you are gonna make a heap of money,'" quipped France.

France then announced that his son, Brian France, would take over personal responsibility of BASCAR (Baseball Association for Stock Car Auto Racing). The younger France announced immediately a series of corporate sponsorships aligned to different teams.

Heretofore, beginning after the All-Star Game, established teams will be referred to as the Old Milwaukee Brewers, the Russell Athletic Supporters, the Houston Astrovans, the Texas Range Rovers, the Hanes White Sox, the Cincinnati Red Dogs and the Thompson Twins. More changes will be announced shortly.

"Those first few were naturals," Brian France explained, adding that Kodak might well be interested in signing on to change the Miami franchise to the Sterling Marlins.

No rules director has been selected, but France said Winston Cup Series Director Gary Nelson had already made a few recommendations.

"Gary and I had lunch this morning with Roger Clemens, and Gary strongly suggested to Roger that he should practice throwing underhand," said France.

Nelson also said that different players would be assigned different bats, based on a computer evaluation of recent performances. For instance, Ken Griffey Jr. would be required to use a bat made of knotted pine. Rafael Belliard, on the other hand, would be issued a new graphite model.

The NASCAR hierarchy also announced that the championship would be decided on the basis of a point system, not wins and losses.

Bill France Jr. was asked if his family had negotiated with the Major League Baseball Players Union.

"I don't see the problem," the elder France replied.

As the Frances, father and son, were spirited off the stage to the strains of Lee Greenwood singing "God Bless the USA," Brian France blurted out, "Union? What union?"

SINCE WHEN DID JACK ROUSH
TAKE OVER NASCAR?
september 1998

LOUDON, N.H.— Many reasons could be cited for NASCAR confiscating the tires of Jeff Gordon's Chevy following Sunday's CMT 300 at New Hampshire International Speedway.

Certainly there were aspects of Gordon's victory that seemed suspicious. With 67 laps to go, Gordon pitted for two tires while most of the other contenders took on four. The shorter pit stop enabled Gordon to take the lead on an afternoon in which he spent most of the day with what was, by Gordon's own admission, "a third- or fourth-place car." What was surprising was that, despite a set of tires more worn than those on the cars of Mark Martin, John Andretti and Dale Jarrett, Gordon maintained the lead, even stretched it out.

No matter what call crew chief Ray Evernham makes, it seems to work. He can change four tires when everyone else changes two, and he wins. He can change two when others change four, and he wins.

Even before the latest conquest, a prominent Ford driver was asked to select one aspect of Gordon's team that he would like for his own team.

"Choose one thing? His last set of tires," the driver responded.

For well over a year, dark suspicions have been voiced about Gordon and his magic tires. Not since the heyday of David Pearson and the Wood Brothers in the 1970s has a driver been so adept at magically going from pretender to contender in the final laps.

Other teams are frustrated. The fans are skeptical. Perhaps NASCAR should have impounded those tires to protect the integrity of the sport.

Integrity is one thing. Confiscating the tires because rival car owner Jack Roush was hopping mad is another.

Of course, NASCAR also took the tires off Mark Martin's Ford, but that was window dressing. In fact, since Roush owns Martin's No. 6, the move made it even more clear that Roush was indirectly calling the shots. No one was suggesting anything fishy (perhaps sticky would be a better word) about Martin's tires.

On site, officials found nothing wrong with Gordon's tires, or Martin's either, for that matter. The rubber then was sent to an "independent laboratory" (forget about NASCAR officials saying where) for testing. As of press time Monday, no further word was forthcoming.

"There's been a lot of talk the last three or four weeks about trick tires or treated tires," NASCAR's Tim Sullivan told the Associated Press. "We don't believe that, but we wanted to prove to ourselves and our competitors that is not the case."

Actually, the "talk" has been going on for much longer. Some have suggested that DuPont, Gordon's primary sponsor, has come up with some sort of magical chemical means to "soak" tires from the inside. "Soaking" is an old ploy, long banned in NASCAR, to soften compounds by applying chemicals. Soft tires grip the pavement better but do not last as long.

"I'll tell you what's in our tires," said Evernham. "Air. Just like everybody else's." (Actually, stock car tires are inflated with nitrogen.)

Publicly, Roush would go no further than to cite statistics about how Gordon's lap speeds seem to improve by "the better part of a second" near the end of virtually every race. But eyewitnesses saw him lobbying fiercely for an investigation in discussions with race officials.

Evernham was just as fierce in denying any wrongdoing. In fact, when he said that Roush "disgusts me," it marked the first public schism in a race for the Winston Cup championship that had been marked by hot competition on the track and syrupy mutual admiration off it.

During a six-week period in which Gordon has won five races, Martin won the other and finished second to Gordon four times.

It's obvious that Gordon and crew have an advantage. The question is whether it is inside or outside of the rules.

If the advantage is within the rules, no one is going to believe it anyway.

IS THERE STILL ROOM FOR SHORT TRACKS?
april 1999

The next two stops on NASCAR's Winston Cup Series are on what are known as "short tracks," which by definition are tracks of less than a mile in length.

Short tracks are important to the sport for several reasons. First, since virtually every driver learns his craft on a local short track, such tracks are important ties both to NASCAR's humble origins and to the humble origins of its heroes. Furthermore, short-track racing is difficult and physically demanding. On a scale of difficulty, surviving 500 laps at Bristol or Martinsville is harder to accomplish than 500 miles at Talladega. Most everyone agrees that the driver's role is proportionally more significant at a short track.

The trouble with short tracks is that most of them are in small towns. The series has recently visited Las Vegas, Atlanta and Dallas-Fort Worth. Sponsors love exposure in those markets. They do not get quite as excited at the opportunity to "wine and dine" their customers in Martinsville, Virginia, where they roll up the sidewalks precisely at 8 p.m.

"I think the short tracks are important, and they're good for the sport," said team owner Bill Davis. "I enjoy them, and I think the fans do, too. How could you not? It's good racing.

"We need to remember our sport came from close fender-to-fender racing and some of the great races from those days. Even in this era, remember some of the great battles like Earnhardt-Wallace and some of the deals those guys had? It helped build our sport and grow our fan base."

But? (A-thousand-one-a-thousand-two . . .)

"The sizes of the markets, from a sponsor's point of view, that's the biggest problem," Davis added. "As our costs go up and we ask for more money per race, they have to justify that. We have to have so much money to make these long trips to the big markets, and that would be a problem for the sponsors. Some of the smaller markets are just not as well suited for what our sponsors are looking for. That doesn't make them bad and doesn't make them undesirable. It's just a fact of life with our sport, or with any professional sport, that big markets are pretty important."

Not all short tracks are in small markets. Richmond, for instance, has a .75-mile track that is one of the sport's model facilities. As long as Bristol can draw 135,000 fans to fill its mammoth seating area, no one is going to suggest that it give up one of its races.

For short tracks to die off, and these factors also come into play at small superspeedways in small towns like Rockingham and Darlington, the sport is going to have to be governed exclusively by money and its pursuit.

That may not have already happened, but it certainly is the trend.

"You have to have a variety of tracks," said John Andretti. "If you're going to go out and run cookie-cutter race tracks everywhere, that's not exciting. What NASCAR provides is a lot of different opportunities for a lot of different kinds of race tracks, a lot of different car setups, and whether you like them or not, the variety gives people a reason to follow you from this race track to that race track. People can see different guys running good. You run the same kinds of race tracks, and you're going to see the same guys up in front all the time."

The same guys winning every race? Maybe guys like, oh, Jeff Gordon and Mark Martin? The devil you say! (In fairness, it should be noted that Gordon and Martin, as well as Jeff Burton, Dale Jarrett and Terry Labonte, are all as exceptional on short tracks as on superspeedways.)

"Sure, tempers flare [on the short tracks]," Andretti said. "You ever sit in a line of construction (on a highway) and, all of a sudden, here comes some guy whipping up in the other lane and he pulls in front of you? You think that makes you very happy? Well, it's the same thing here. When a guy takes away something you rightfully earned, your temper can flare and rightfully so. Plus, it's a pretty emotional sport to begin with.

"If they built a short track in Seattle—well, maybe Seattle wouldn't be the best example with all the rain that's plagued us this year—but if they put up a short track in a big market in some place it doesn't rain all year and did it right, we should take a look at going there. Richmond is a great example of a great job of building a race track. That would be super, in a major market, and I think people would pack the place. You go to Richmond, and you see a lot of great racing, great side-by-side racing. Those types of tracks would work well anywhere."

"Do we need [short tracks]? Yeah, we need them," added Andretti's crew chief, Robbie Loomis. "We need at least a few of them. Do we need as many as we have? I don't know if we need that many, but we need some. They bring a lot of excitement. They bring a lot of competition."

Car owner Michael Kranefuss offered a balanced view of the whole dynamic.

"You can look at it from a historic point of view, and short tracks definitely should be there," he said. "There may be one or two facilities that haven't been able to keep up with what you need to have in terms of amenities or support for the competitors . . . but they lend a lot of excitement, and I think they should be part of the schedule.

"You can look at it from a marketing point of view and say, 'Are there places where we would rather be?' Everybody's opinion is as good as the next guy's. I think NASCAR's position is to add rather than take one away here or there, so I think we're going to keep the short tracks. Is a race in Chicago more preferable than one in Bristol from a marketing standpoint? No doubt. But from that same standpoint, a short track in Chicago would, from a marketing standpoint, be more preferable than a large superspeedway in Bristol. It has nothing to do with the size of the track or even, for the most part, the numbers sitting in the grandstands."

Of course, in terms of advertising products and providing exposure for sponsors, it is pertinent to note that all the races are on television. Do the viewers care whether a race is run in Vegas or Martinsville? Are they more likely to buy a product adorned on the winning car? Probably not, and certainly not much.

"The marketing aspects of the short tracks, that's sort of a business decision," said Kranefuss. "I think they should remain part of the schedule, at least the better facilities, and I think you have to take advantage of the popularity of the series that demands you to open new markets. Between those two, you'll have to build a prudent plan."

Driver Ward Burton grew up not far from Martinsville.

"We're a sport, but we're entertainment too," he said. "If we're going to race, we need people in the grandstands and people watching us on television. The whole thing behind entertainment has always been, 'Give the people what they want,' and they want short-track racing. I don't think they want it every single week any more than they want to see any single type of race track every week, but I think short track racing is important to them.

"It's important to the drivers, too. Just about every one of us started on the short tracks, and it's where we came from. It's not exactly a throwback to those days because the cars are so different, but it's a nice feeling. . . . I think a lot of other forms of racing are missing out by not being able to run short tracks, and it's something NASCAR has on a major league level that nobody else does."

"I'd hate to see the short tracks die off," said Andretti. "Surely they can put enough seats around those places to keep going. The racing is super, and the fans really seem to love it. Good competition and the fans being happy . . . why would you mess with that?"

RULE BOOK? WHAT RULE BOOK?
june 1998

SONOMA, CALIF.— Winston Cup stock car racing is America's most unpredictable sport.

Not the racing, mind you. This year I've seen the Texas Rangers score 10 runs with two out in the seventh inning. I've seen an Oakland catcher fake a throw and hit himself in the foot. In NASCAR, they *shoot* themselves in the foot, and usually it happens high above the track in an air-conditioned booth. It's the officials I'm talking about.

I never know what's going to happen.

At Richmond, on June 6, with six laps to go, a race was stopped "in the interest of competition." As a result, one driver who had been sure to win, Dale Jarrett, did not, and the fans got to see a rousing finish in which another, Terry Labonte, bumped his way past Jarrett as the two sped to the start-finish line. The means by which this spine-tingling finish was provided? The red flag, which had never, in all the annals of international motorsport, been used for such a purpose.

Here at Sears Point Raceway on Sunday, a Ford driven by Jeff Burton crashed into a trackside barrier so hard that the barrier had to be repaired. This was a classic circumstance for a red flag to be waved and the competition stopped. Instead, only the yellow caution flag was waved.

"Fixing the barrier will only take a lap or two," said NASCAR spokesman Jeff Motley. "There's no reason to put out a red flag."

For 11 laps—on a road course, 11 laps under caution takes a very long time—the cars crawled around the 1.95-mile layout, drivers and teams bewildered, wondering what to make of this most recent NASCAR rewriting of the rules. Eventually they all pitted, and the race resumed.

Did this maneuver have any effect on the outcome? Well, Jeff Gordon seemed to have the superior car, and eventually Gordon won. But at least one driver, Sterling Marlin, felt robbed. Marlin thought the yellow flag robbed him of a superior strategic position.

"The fans paid to come see a race," steamed Marlin. "They didn't pay to see us ride around under caution. Our strategy went out the window."

Translation: Marlin wound up finishing seventh.

How can I best describe the difference between a NASCAR-sanctioned event and one held by the National Football League or the Pahokie Industrial Softball Federation? NASCAR has rules, but they are not allowed to get in the way of what NASCAR wants to do, which ostensibly—these are their words, not Sterling Marlin's—is to provide an exciting spectacle.

If a baseball player bunts, and the ball bounces into his thigh as he leaves the batter's box, he didn't mean for it to happen, but he is still out. In NASCAR, if something comparable to this happens, he may be out, and then again, he may be safe. It depends on who he is, how many fans he has and what his sponsor is. NASCAR officials always know what is important.

I do not believe these men out-and-out rig the races. By their own strange definitions, they are acting in what they believe to be the best interests of the sport. But if you're sitting in the grandstands at Richmond, and your favorite driver just happens to be Dale Jarrett, you leave that night thinking you —and Dale—got robbed.

NASCAR desperately wants the attention focused on what it calls "the stick-and-ball sports": baseball, football and basketball. The Winston Cup Series has the numbers—television, attendance, commercial involvement—to warrant such coverage.

But when the New York Times shows up at your doorstep, you can't wave a red flag for no good reason one week and withhold that flag for every good reason a few weeks later. They will laugh at you. You are confirming their suspicions. They will not understand.

PATCHES ON THE UNIFORMS? WHY NOT?

april 1999

While stock car racing is a newcomer to the ranks of America's sporting mainstream, in some ways the Rapid Roys of NASCAR are ahead of the game.

Take commercialism. Please.

Rumor has it that major league baseball is thinking about allowing its teams to place advertising patches on the sleeves of the players' uniforms.

Race-car drivers have been wearing advertisements on their uniforms, not to mention their cars, since the sport began. From the moment some grease-stained short tracker figured out it was going to take more than 25 last-place dollars to replace the crumpled heap of metal he destroyed on the opening lap, racers have enthusiastically sold practically anything in pursuit of advertising bucks.

It should surprise no one that, when the taboo subject of putting a Kahn's Weiners patch on the sleeves of Yankee pinstripes is raised, the typical answer in NASCAR is, "So what?" If there was money to be made from taking off one's shirt, the average stock car racer would gladly tattoo Goodyear or Chevy across his chest.

"My first thought is, 'What took them so long?'" said Ward Burton. "Just about every professional sport has at least some corporate involvement, and I don't see why baseball would be any different. If they do it now, it won't be long until football and everything else will be doing something along the same lines. The only sport I see having a problem with it is basketball, and that's only because they don't have sleeves."

But what about dignity, you ask? Dignity, shmignity. As Lee Petty might opine, "Dignity don't pay the light and water." Not to mention the gas and oil.

"Undoubtedly we will soon be hearing about the destruction of the 'purity' of baseball, and that's not necessarily talk I find bad," added team owner Michael Kranefuss. "It's not that I care about the 'purity' of one sport as much as I feel those types of marketing dollars will eventually find their way to a place that wants them, and NASCAR is certainly a place like that. All sports have taken corporate sponsorships; motorsports is the only one that has made it a partnership with the corporation."

To summarize, purity is one thing, surety is another.

"I don't follow baseball a lot," said Kranefuss' driver, Jeremy Mayfield, "but I can remember seeing pictures from some of the old ball parks, pictures from 60 or 70 years ago, long before there was a NASCAR. On the outfield fences you see advertisements painted. There weren't any painted-up stock cars then, at least none on a major-league level, but major-league baseball used what it had to do to stay afloat.

"There isn't anything wrong with that at all, but I think it shows that using the ball parks and the players isn't something that started with motorsports. Maybe we just do it better."

Ever greater commercialism is inevitable. With all due respect to the self-selling competitors of NASCAR, I don't want to see the day when the Boston Hanes Sox take on the Houston Astroturfs in the MCI WorldComm Series.

Yet, I don't have a problem with NASCAR. Where's the consistency?

Darned if I know. I just don't mind it in NASCAR, and I'd mind it a whole lot in baseball. Call me critical. Call me hypocritical. I'm just accustomed to hearing race-car drivers say, "I couldn't have done it without that stout motor in the Goody's Headache Powders/RC Cola/Bosch Platinum Spark Plugs Pontiac."

Heck, I just cut the sponsor names out of the quote. So sue me.

MIRACLES HAPPEN, AND NOT ALWAYS FOR THE BEST
november 1999

Auto racing can be the setting of the most wondrous, and also the most awful, coincidences.

Wildly improbable coincidences have marked the history of NASCAR. Richard Petty's entire career consisted of one unpredictable godsend after another. Petty won his 200th race on the occasion of President Ronald Reagan's only visit to a race. His first championship occurred during the year in which the sport's leading driver, Glenn "Fireball" Roberts, died in a fiery crash. Petty's final championship preceded the first of Dale Earnhardt, who ended up tying his record of seven. Petty's final race, in 1992, just happened to be Jeff Gordon's first.

But the magic is sometimes black. The careers of drivers like Davey Allison, Tim Richmond and Ernie Irvan seemed somehow ill-fated from the very outset.

And then there was Neil Bonnett.

Few drivers have ever been so well regarded by their colleagues as Bonnett, an amiable Alabaman whose career was apparently cut short when he suffered serious head injuries in a 1990 crash at Darlington. In the aftermath of that crash, it was widely reported that doctors had advised him never to race again.

Bonnett, taking advantage of his engaging personality, moved effortlessly into the broadcast booth, quickly becoming more popular there than he had been as a competitor. Bonnett never lost the racer's itch, however. Even as his broadcasting career took off, Bonnett began dabbling with the idea of driving again. He became a valuable test driver for Richard Childress Racing and his fishing buddy Earnhardt. In 1993, Bonnett decided that he wanted to race again, and eventually he found a doctor who would certify his competence to compete again. In gratitude, Childress entered a second Chevrolet for Bonnett in the DieHard 500, held on July 25, 1993, five days before Bonnett's 47th birthday.

That was my first visit to Talladega Superspeedway, the fastest oval on earth. Talladega is scary enough because of its speed and the closeness of competition, but the track has also gained a mystical aura over the years because of weird tragedies that occurred there. It was where a promising young driver named Larry Smith died after what had seemed a routine bump against the wall, where Bobby Isaac once pulled into the pits claiming a mysterious voice had told him to get out of his car, and where the beloved DeWayne "Tiny" Lund met his maker in a back-stretch pileup. Until that day, I had never seen tragedy, at least not in motorsports, firsthand. A crash occurred that day in which a car driven by Jimmy Horton sailed over another car and out of the entire facility. Horton suffered only minor injuries, but another driver named Stanley Smith was almost killed, his career ended, in the same wreck.

In a separate incident, Bonnett, with all eyes monitoring his comeback, flipped his No. 31 Chevrolet crazily in the tri-oval section of the huge track. Scarcely a word was spoken in the press box as emergency workers arrived on the scene. Miraculously, Bonnett climbed out of the smoking heap uninjured.

Almost everyone present had the same reaction: Oh, thank God he's OK, and surely this nightmarish crash will get racing out of his system. It didn't. As we all listened incredulously, Bonnett told a national audience of both radio and television listeners that he had every intention of continuing his comeback. It was just another crash, he said, just a part of racing.

I can't speak for everyone, but I remember distinctly my own thoughts: If ever a man received fair warning . . .

Bonnett ran one more race, but it was nothing much to speak of. Childress installed him in the season finale at Atlanta merely to help nail down Earnhardt's sixth championship. Earnhardt needed only to finish in the top 35 or so to win the title, so Bonnett went out and ran just a handful of laps before parking his car, thus giving Earnhardt one more car he didn't have to beat.

A short while later, Bonnett announced plans to run a limited schedule of events in 1994, driving a garish lemon-and-pink Chevrolet carrying No. 51. Before he ever ran another competitive lap, Bonnett was killed in the new car, testing for the Daytona 500, on February 11, 1994. At the time of the fatal crash, I was driving down Interstate 95, en route to the track from the Carolinas. When word of the tragedy arrived via a radio report, my reaction mirrored that of a million others. I wept quietly and pulled off the highway to compose myself. Unlike most other mourners, I then placed a call to the office of the newspaper where I worked at the time.

The sports editor, who already knew of Bonnett's death, said, "What was he even doing in a race car? He was good on TV. He had another career. He didn't have to drive."

"No, I think he did," I said. "You've got to remember something about these guys. They're crazy to begin with. Who would ever want to strap himself into a race car and drive wide-open anyway? They're fundamentally different from the rest of us."

The next time a fatality occurred, I wasn't so lucky.

In October 1995, I was cranking out copy in the Charlotte (now Lowe's) Motor Speedway press room while a Sportsman race was going on at the track. This series had been devised as a means for short-track drivers to get experience racing on high-banked superspeedways. The only time such a race merited much coverage was when someone was injured in one of the crashes that inevitably occurred when inexperienced drivers were learning their trade. The race was, however, being shown on television monitors in the infield media center, and I was paying partial attention.

Moments before, a crash had occurred when a skidding car crashed headlong into the water-filled plastic barrels that guarded the entrance to pit road. While flames smoldered in a another car stricken nearby, a gusher of water splashed high into the air as the first car crashed into the barrels. Partly because we are naturally cynical and partly as a reflex, most journalists have cultivated a gallows humor over the years. As I watched track personnel mop up the mess, I cracked, "This has got to be the only series where one guy could burn to death and another could drown in the same wreck."

Within minutes, I regretted that little quip.

I happened to be looking at the television screen, having just completed a story on Winston Cup qualifying or some such. Shortly after the race was restarted, two cars crashed in front of a bunch of others. A third driver swerved radically to the right, failing to notice that he was veering into a fourth car. It was almost as if the third car burrowed under the left side of the fourth, turning the fourth car sideways on the banking. With showers of sparks flying everywhere, the fourth car hit the outside wall top first, pinning it (the fourth car) against the concrete.

The impact sheared off the entire roof of the car, sheet-metal, roll-cage, everything.

With the television monitor affording a frontal view, the car drifted lazily back off the wall, landing back on its four wheels. That's when it became sickeningly apparent to everyone that the driver of the car, a young man named Russell Phillips, had been beheaded. Several days later, while editing a weekly trade paper, I looked at a series of photographs of the crash and made a decision to print a sequence in black-and-white, not color, because the photos showed a stream of what looked like gray smoke in black-and-white. In color, the stream was pinkish because it emanated from the disintegration of human tissue. The fence, the wall and the track surface had to be sanitized.

Few things are worse for a journalist than having to write about such a grisly scene. There is the horrible sight of grieving family members and acquaintances at the infield hospital, the wait for official notification of that which is sickeningly obvious, and the actual act of sitting down, submerging one's revulsion and writing about it in words carefully and sadly chosen.

Russell Phillips was no bright star on the horizon, no newcomer who had already made a small fortune on his way to the big time. He was a young man with a family, a guy who gave his time to work with youth groups. He had been a fireman who raced only as an avocation. He had been tragically brought to that time and that place by a simple desire to give it a whirl, to see what it was like racing at the high speeds of a big asphalt palace. In those days, my job consisted of more than just covering auto racing, and I was thankful that I could not spend much more time at the speedway that fateful afternoon because I had to cover a high-school football game. I almost had a wreck, too, because just as I was exiting from one interstate highway, 85, to another, 77, my car swerved. It was because the shock had worn off, and I started retching. Of course, many of these details were not written at the time, and perhaps should not be now. The more time passes, though, I think that the gory details should be known, that they should be printed and that they should act as some sort of safety valve against something like that ever happening again.

There were no more Sportsman races at the track. In the more than four years since, I have seen nothing to which it can be compared. Thankfully, when in 1999, spectators were killed and injured in an Indy-car race at the track, I was far, far away, covering a NASCAR race in California.

Dale Jarrett at an event in Phoenix before the Checker Auto Parts/Dura Lube 500, November 1999. *Tyson Cartwright*

• • • • • • • • • • • • • • • **chapter 5**

COLOR

WHY STOP WITH NAMING A RACE TRACK?
february 1999

DAYTONA BEACH, FLA.— As of March 1, I intend to change my name to Cheez-It Keebler. Cheez-It, you may recall, is a brand name of a snack food, and gosh knows, I like snack foods. Keebler is Cheez-It's parent company, and I have decided to sell my journalistic career to Keebler.

In exchange for a yearly salary, beginning at the previously noted date, all of my stories will be bylined:

BY CHEEZ-IT KEEBLER
Gazette Sports Reporter

In addition, my career will be used to draw attention to other fine Keebler products. For instance, Larry Pearson of Spartanburg, South Carolina, drives the Cheez-It Pontiac in the NASCAR Busch Series, Grand National Division (BGN). Before each BGN event, part of my job as Cheez-It Keebler, *Gazette* Sports Reporter, will be to keep readers informed about the exploits of Larry Pearson, Cheez-It Driver. I will write—free of charge—everything I can possibly think of that is good and wholesome about Larry Pearson for *Gazette* readers.

As a matter of fact, negotiations are under way to make this newspaper, the *Gaston Gazette*, the *Diet Sundrop Gazette*. What we now call the "sports" section will soon be known as the "Tony's Ice Cream" section. The news section will become the "Remington Firearms" section. This may cause a problem for awhile, but soon our readers will equate sports with Tony's Ice Cream, which, after all, is a wonderful snack suitable for almost any recreational/sports activity. Soon the readership will equate crime with Remington, which nonetheless encourages all Americans to own licensed firearms.

By dropping the reference to Gaston County, which after all, does not pay any rights fees, the *Gazette* can offer its information-disseminating services more economically. In the end, it will be better for everyone.

To answer critics who question the objectivity of the *Diet Sundrop Gazette*, I sought out the *Gazette*'s managing editor, Quakerstate Pennzoil. Quake, as we refer to him around the newsroom, assured that we would have no problem as long as I stayed clear of Cheese Nips, Mountain Dew, Haagen Dazs and Winchester. These products had every chance to align themselves with the *Gazette*, but it was a fair fight and they lost out in negotiations. After all, business is business.

This column has absolutely nothing to do with the possible consequences and loss of dignity stemming from the renaming of Charlotte Motor Speedway as Lowe's Motor Speedway. It is merely a reflection of where America is headed and a desire by yours truly to be on the cutting edge.

I'm kidding.

VEGAS IS A FREE-SIN ZONE
march 1999

LAS VEGAS, NEV.— Want to know the best thing about Vegas?

It's a free-sin zone. No, really, it is.

NASCAR's Winston Cup Series is loaded down with those who would take up arms in defense of morality, religion, family values and social piety. The Motor Racing Outreach has a 50-foot tractor-trailer rig parked out in the infield, with satellite uplinks and live hookups to contact anyone who cannot be reached by old-fashioned prayer.

To hear these people talk, in order to coordinate their track activities they need a high-tech apparatus that would make NASA, NATO and the Strategic Air Command proud. If they ever stop saying, "Gentlemen, start your engines" at the start of these races, they ought to try, "Onward, Christian Soldiers."

But last week morality drew a bye. The Lord takes a break in Vegas. It's a free-sin zone.

A tour of the casinos reveals a veritable who's who of the Christian, the conservative and those all too fond of prescribing just how everyone else should live his or her life. But they will elbow you out of the way to secure a preferred place at the craps table or the roulette wheel or in the poker room.

"Land o' Goshen, child, step aside. There's money to be won!"

Two years in a row I have ridden across this great country of ours in an airplane stacked with NASCAR mechanics, drivers, car owners and media types. Each was telling stories of last year when he hit the slots for $10,000, or the year before that, when he and his buddies flew up after Phoenix and made a killing playing blackjack, or when he took home a cool thou betting Phoenix against the Lakers and taking the points.

I have yet to hear the first story of anyone losing his shirt in Vegas. Apparently, not only is it a free-sin zone, it is a place where everyone wins. Isn't this amazing? How could they possibly manage to erect an Egyptian pyramid disguised as a hotel, or scale replicas of the Eiffel Tower, Empire State Building and Statue of Liberty, all the while giving away money by the wheelbarrow load?

This is a bigger story than Santa Claus.

I made the mistake of referring to this concept in the presence of one of the NASCAR apostles.

"Where does it say in the Bible that you can't gamble?" he roared.

"Umm, I believe it has something to do with Jesus throwing the moneychangers out of the temple," I replied meekly, making no claims to being erudite in matters biblical.

"We ain't gambling in no temple, son. This is a casino."

Praise the Lord. Amen, brother.

There is one corollary to the secret zone where sinning is hunky-dory. It's the self-fulfilling prophecy. One has to believe to be successful.

This I learned from personal experience. While I am not much of a gambler, I almost always lose. I come to Vegas, set aside my $50 and proceed to lose it with deceptive alacrity.

Alas, ye of little faith, it is not enough merely to gamble. One must gamble righteously.

RULES TO COVER NASCAR BY
february 1999

DAYTONA BEACH, FLA.— As much as I hate to admit it, the season is upon us, so it's time to lay some ground rules.

For all of you people who insist on calling or writing me to say that I don't know what I'm talking about: First, I'm not talking; I'm writing. Second, here is what I'm writing about.

Thanks to Messrs. Clinton and Starr, and Mlle. Lewinsky, we've all grown nauseatingly knowledgeable in the rules of parliamentary order. So, for the convenience of all of the folks who've had the impeachment proceedings get in the way of the *Jerry Springer Show* lately, here are this year's parliamentary points of information from the NASCAR beat:

· I do NOT have a favorite driver. Many fans are fond of drawing me close and whispering, "C'mon, tell me who *your* favorite driver is. I mean, you can tell me."

If I were a fan, I would have a favorite driver. However, I am a reporter, and as such, I pull for *stories*, not *drivers*. I like Jeff Gordon, but after he has won three races in a row, I'm tired of writing about him.

· I do not like *writing*. What I love is being a *sportswriter*. The two are different. What a sportswriter does is go out to dinner and share embarrassing, raucous stories about everyone who is not there to defend himself or herself. The price I pay for being able to do this is that I must write stories. It's worth it.

· Do not bother to accuse me of being a smart alec. It is what I do for a living.

· Do not bother to accuse me of not telling the truth. A writer's job is not to tell the truth, and it never has been. What we do is tell what people tell us the truth is, and that is quite a different matter. On a good day, we get close to the truth, but if we really want to write the truth, we have to do it in the form of fiction. That way we can't be sued. In many ways, the truth hurts.

· It is very difficult for you to insult me, simply because I know that you will read into my stories what you will. I have received calls within a 10-minute period, one calling me pro-Ford and the other pro-Chevy, one calling me pro-Earnhardt and the other calling me anti-Earnhardt, about the *same story*. Here is the message I get from such a phenomenon: the people who called me must, by definition, be *nuts*.

· Actually, I have a healthy attitude about criticism. Basically, I shoot for a situation in which about 15–20 percent of the readers hate what I write. If I don't get that 15–20 percent hatred rate, then I must be putting the readers to sleep. No one reads my stories more voraciously than the readers who hate them. Hatred sells papers.

· I couldn't care less about other people's opinions of what I write. I try to be fair. By whose standards? Mine. It being a free country, I can't control what others want, what they love and what they hate. By eliminating those considerations from the whole process, I save myself untold stress and aggravation.

Now, on to specifics:

• I like the driver of the No. 60 Power Team Chevrolet, and because of this, I will continue to call him Geoff Bodine. By insisting that he now be called Geoffrey, my friend Mr. Bodine is exposing himself to untold ridicule, and I refuse to be a part of it. Geoffrey, indeed. The only way he could possibly have done worse would be to have insisted on being called Little Lord Fauntleroy Bodine.

• I hope that, for the third year in a row, something goes wrong at Texas. I really get tired of hearing other writers say, "Oh, gosh, that Texas race was a nightmare." It wasn't a nightmare. I had plenty to write about, and that's what I want when I visit a city.

• Here is what I live for: to take people who make more money in a week than I will in my whole life and to ridicule them. At Texas Motor Speedway, and in the state of Texas in general, it is so easy.

• Will I knuckle under and call "that track in Concord" Lowe's Motor Speedway? I don't see any way of getting around it. I don't see pro-football writers balking at Ericsson Stadium. But, I don't like it, and I won't like it until I see evidence that the $35 million over 10 years is being used to make racing more economical and/or better for the average fan. Yeah, like that's going to happen . . .

Maybe I'll just be sarcastic and refer to it as "Lowe's Motor Speedway, named after a hideous home-improvement chain." What's worse, this is only the leading edge. I predict that within my lifetime, America's most famous sports team will be named something like the General Motors Yankees. They might even be called the Fabulous Pontiac Firebirds. I hope they at least continue to wear pinstripes. Imagine a World Series between the Fabulous Pontiac Firebirds and the Jimmy Dean Sausages. It just don't get no better.

· Every year I predict that Ken Schrader and Bill Elliott will win another race. I hope they do, but I no longer believe it myself. Every year I predict rookies will win races. I hope they do, but I no longer believe it myself. Every year I predict someone other than Jeff Gordon will win the Winston Cup championship. I hope someone does, but I no longer believe it myself.

In short, what I am trying to say is: NOTHING EVER CHANGES.

LOST AMIDST THE DEARBORN ENGINEERS
august 1998

DEARBORN, MICH.— As Bill Goodrich, who broadcast Clemson University football games when I was about 6 years old, used to say when someone like Bo Ruffner would score a touchdown, "Whew, mercy!"

On Thursday, I attended a press symposium hosted by the Ford Motor Company to publicize its motorsports program, and I was repeating those words as I staggered out.

How can I describe this? You've heard people say, "You don't have to be a rocket scientist to understand this"? Well, I would have had to have been a rocket scientist to grasp fully what was being described.

I know just enough to be dangerous, at least where the technical side of stock car racing is concerned. I know what a carburetor restrictor plate is. I even know where it is located. I know what a driver means when he says his car is "loose." I know the difference between a Ford and a Chevy and that almost everything that happens in NASCAR is "one of them deals."

But for close to five hours, I heard engineers and aerodynamicists discuss the finer points of equations and theorems and acronyms. You know, "We used our CNCs to subdivide the BFQs into a manageable vortex of FMAs and MLDs. Once the problem was resolved, we realized a measurable enhancement of the kinematic yaw at optimum speed."

How does the average liberal arts graduate respond to such jargon?

"Oh, yeah, well, I once read a book by Dostoevsky. And till you do that, don't talk to me about drag coefficients"?

Of course, these guys were so smart they could do that, too.

"Mr. Dutton, I must admit that I haven't a comprehensive knowledge of Dostoevsky, but if you would allow me, I was particularly stimulated by a recent reading of *The Possessed.*"

Sheesh.

Ford uses a computer called a Cray. Apparently, what Everest is to mountains, Cray is to computers. The engineers—they had been dressed in snazzy "Ford Racing" golf shirts to soften the geek image—talked about submitting data that took hours for the Cray to run.

They showed us computer-generated illustrations of the amount of air pressure exerted on each square inch of the Taurus' surface. One engineer estimated that the leveling off of a section of the Taurus, a section not covered by templates, could translate into a gain of 10 horsepower or its practical equivalent.

Then we went to a dynamometer, but not just any dyno. This one was capable of simulating a race track, or even simulating a certain driver on that race track. They could adjust the humidity and the temperature. You could conceivably yell out, "OK, put Indianapolis Motor Speedway in Alaska on Christmas, put Emerson Fittipaldi behind the wheel of a Jack Roush Taurus, and run 300 laps," and the operator could reply, "Yo, man, no problem."

Everyone understands that racing is highly complex and that computer technology has revolutionized the way it is conducted. But I'm here to tell you, no matter how much you think you understand this, you are understating the case. For the first time I understand why an engine that could generate 600 horsepower 10 years ago is now closing in on 750.

I don't care whether you're talking foot-pounds of torque or pounds of downforce per square foot, the possibilities are endless in this computer-driven age.

LET'S PLAY THE NAME GAME
february 1998

Maybe I'm getting old, but lately I've found myself confusing similar names. Recently I meant to make a comment about the basketball star Chris Webber and instead called him Dick Weber. Dick Weber was a bowler, a lefty as I recall.

My mind has started playing word association games without my permission!

Imagine the possibilities.

For instance, it would be quite a stretch to confuse road racer Keke Rosberg with fatherly golf commentator Bob Rosburg, but no more so than to confuse old TV character Matt Dillon with young actor Matt Dillon. The former, played by James Arness, was often aided in his duties by Festus, played by Ken Curtis. Ken was no relation to 1970s major league baseball pitcher John Curtis. Or to actor Tony Curtis, either. Tony Curtis did, however, manage to get out of a lot of tight jams when he portrayed Harry Houdini in a movie that nonetheless could never be described as a "whodunnit."

Richard Petty is the son of Lee and father of Kyle, but he is no relation to Norman Petty, who was Buddy Holly's manager. Neither Buddy nor Norman was ever a member of the Hollies, however.

Michael Jordan lives life on a grand scale, as did Jordan "Bick" Benedict, the main character in the movie *Giant*. Bick Benedict was played by Rock Hudson, who had nothing in common with Charles Hudson, the pitcher, or Henry Hudson, the explorer. Ralph "Shug" Jordan coached football at Auburn. Henry Jordan played for the Green Bay Packers, Lee Roy Jordan for the Dallas Cowboys. They, too, were unrelated.

Race-car drivers Jeff and Robby Gordon are unrelated. So are Derrike and Mike Cope. West Coast driver Ernie Cope is, however, Derrike's cousin. David, Ricky and Larry Pearson are all from the same family, but ex-football stars Drew and Preston Pearson are not.

Cale Yarborough was not related to the late Lee Roy Yarbrough, but they were NASCAR stars at the same time. Cale

Gundy played quarterback at Oklahoma. Lee Roy Selmon also played for the Sooners.

Gianpiero Moretti was one of the winning drivers in the Rolex 24 at Daytona Sunday, but he bears no connection with Laura Maready (roughly the same pronunciation), who works in public relations at the Texas Motor Speedway.

Neither Bobby, Al, nor Al Unser Jr. has anything in common except the last name with ex-Phillies outfielder Del Unser. Nor does Kenny Irwin, the new driver of Robert Yates' No. 28, even know senior golfer Hale Irwin. Or movie producer Irwin Allen. Robert Yates is unrelated to Rowdy Yates, a character portrayed by Clint Eastwood on the old television series *Rawhide*. The star of that show was the late Eric Fleming, who apparently never knew of the existence of Olympic figure skater Peggy Fleming or Fleming Thornton, my father's high-school football coach.

Banned baseball star Pete Rose bears no relation to Charlie Rose, but I could see Pete as a guest on Charlie's TV show. Bobby Rahal drives a mean Indy car, but I doubt he could turn a double play as well as Cookie Rojas. He might, however, be capable of pitching as well as Mel Rojas did for the Cubs last year.

Spencer Tracy, the actor, knew nothing about Paul Tracy, the racer, or Jim Tracy, the outfielder. He may have read *Dick Tracy* occasionally. Bruton Smith, the CEO of Speedway Motorsports, is unrelated to Rankin Smith, who owned the Atlanta Falcons.

Bruton, however, does bear a remarkable likeness to Cogswell, of Cogswell Cogs, who was Mr. Spacely's rival on *The Jetsons*.

Hmm. I am in need of a stopping place . . . got it.

Carl Yastrzemski.

I'M MORE THAN JUST A GEAR HEAD, AWRIGHT?

october 1997

It's a fairly common occurrence when I'm hanging around at the office, not that I live there, mind you. Everyone knows that I have established residency at the Commonwealth of HoJo, like our other beat reporters. I'm running as a delegate to the next national political convention from Americans Out of State.

"Madame chairperson, the Great Out of State awards its Frequent Flyer Miles to the next President of the United States, the man who brought Pita Stuffs to the drive-through windows of the land, the honorable Dave Thomas!"

Oh, yeah, the common occurrence.

Some high-school coach happens by to see Darin Gantt, our personable high-school sports editor. It happens all the time. I, on the other hand, never get a visit from Chad Little. I just see him in airports.

So I introduce myself to the coach. We dress somewhat alike. He's got "Ashbrook Soccer" on his golf shirt; I've got "Slim Jim Racing" on mine.

His expression droops. "Oh, yeah, you're the NASCAR guy."

Pretty soon, we drop any pretense of conversation. He waits for Darin to show up. I go back to filing expense reports and surfing the Internet for "www.goracing.com."

Now I proudly admit to being an unabashed lover of motorsports. I went to my first NASCAR race when I was 7 years old. "Gentleman" Ned Jarrett won it, and I had to sleep in the back seat of a 1964 Plymouth the night before. I remember being more than a little ticked off that "Fearless Freddie" Lorenzen couldn't run down Mario Andretti at the 1967 Daytona 500. That's kind of the equivalent of knowing the words to every song Hank Williams ever wrote. (Oh, by the way, I can do that, too.)

But just because my job is racing, that doesn't mean I stopped being a fan of all the other sports.

For the record:

- My all-time favorite football player is Tom Matte.

- My all-time favorite baseball player is Carl Yastrzemski.

- My all-time favorite basketball player is Jerry West.

- My all-time favorite hockey player is Eddie Giacomin. Not really, but aren't you impressed that I can spell it?

- I was at the game on April 8, 1974, when Henry Aaron belted his 715th home run. It was my birthday. My brother's birthday, February 5, is the same as Aaron's. John Hiller, Catfish Hunter, Gary Carter and Larvell "Sugar Bear" Blanks were also born on April 8, as was the stock car driver Robert Pressley.

- I played on a state championship high-school football team. Not notably, but still . . .

- I wrote a book about high school football.

- I have attended six major league baseball games this season, none of them in Atlanta.

- I was once the sports information director of a small college.

- Last year, on an off weekend, I attended the Texas-Oklahoma game in the Cotton Bowl.

Man, I got credentials, and not just to the UAW-GM Quality 500.

Smoke billowing from Jeff Gordon's tires after he was bumped and spun in a circle during the 1997 Goody's 500 at Martinsville Speedway, Martinsville, Virginia. Amazingly, Gordon went on to win the race. *Tom Whitmore*

SCENES

RACING THE OLD-FASHIONED WAY
april 1998

HICKORY, N.C.— A Winston Cup off weekend gave me a chance to give my sister's three children a taste of racing the old-fashioned way with a trip to Hickory Motor Speedway's Galaxy 300.

I adore the annual Busch Grand National race here and would have been honored to cover it. Writing about Hickory is a great assignment. Very little is spoon-fed. When you hit the pavement after the checkered flag falls, you make a deal with another writer; he goes up pit road and you go down it, then you both get all the information you can and compare notes. As the drivers climb out of their mounts, most of them have a bone to pick with one of their colleagues. It makes for colorful writing, but this was a chance to give the kids a glimpse of real racing, so we bought our tickets, and I passed up the play-by-play.

The boys' love of the sport has grown in recent years, in no small part due to the fact that their uncle makes his living writing about the larger-than-life heroes of NASCAR's Winston

Cup Series. It has been a sore spot to them that they could not go to the races with me.

"Unc, why can't you take me to the races?"

"I wish I could, Jake, but I have to work, and there's nobody to look after you and your brothers."

The "that's-not-fair" look has grown all too familiar. Adults know that the world is patently unfair, but the young should be shielded from that realization for as long as possible.

Free weekends are rare, but I have tried to make the best of them. Exactly a year ago, the boys and I drove to Cherokee Speedway in Gaffney, South Carolina, where they tasted the manly art of dirt-track racing. In fact, they tasted more than a little of the actual dirt. It was a cool night, and Vince, then not yet 7, was a bit underdressed. The poor kid wrapped himself up in my jacket and went to sleep on the hard concrete, hands pressed over his ears, even as the late models were sliding around the clay at breakneck speeds.

This year the boys made the step up to Busch Grand National.

They regarded me with disbelief when I explained to them that Hickory Motor Speedway was actually smaller in size than the Gaffney half-mile. The concept of bigger not always being better is alien to the average 10-year-old. But then I went on to say that, since the track would be crowded, occasional crashes would be almost unavoidable.

"You mean them cars will be knocking the crap out of each other?" Jake asked.

"Umm, basically, yeah."

"Cool." His brothers nodded enthusiastically.

We took a cooler full of Pepsi and Sprite, one container of fried chicken breasts and another of legs, Mississippi mud cake, about a dozen small bags of potato chips, deviled eggs, potato salad, and ham-and-cheese sandwiches. Eating out of the trunk of my Dodge, high on a grassy hill above the third and fourth turns, we feasted with all the enthusiasm of Pilgrims freshly landed at Plymouth Rock.

"Durn, Grandma sure knows how to fry chicken," said Jake. "This is way better than Kentucky Fried."

Vince quickly recovered from the mild case of car sickness

that had reared its ugly head—or perhaps dipped its nause-ated one—on the way up. Rolling down the windows for fresh air had helped avoid a messy scene, and now the embattled youngest was fit and ready to scrap with his brothers again.

"Gimme another chicken leg, Uncle," he said. "But make sure it ain't got no burnt place on it."

"The chicken ain't burnt," said Ray, rising to Grandma's de-fense. "You're just retarded."

Kids don't show a lot of sensitivity when they're busy in-sulting each other.

Soon we were nestled into the wooden splendor of the Ralph Earnhardt Grandstand, wedged in with all manner of ice chest-carrying folk. I had to make only one minor seating adjustment. Ray and Jake agree on nothing. As an example, Ray is a Jeff Gordon fan; Jake is a Dale Earnhardt fan, and as all other fans realize, never the twain shall meet. After the first minor scrap, I calmly reached over, uprooted Vince and stuffed him back into the space between his two brothers, thereby bringing relative peace and tranquility back to our happy troop.

I hope I was able to impart to the youngsters the simple wonder of short-track racing. Life is too simplified for kids to-day. They play their baseball on Nintendo sets instead of sand-lots, watch movies via the VCR, not the silver screen, and see racing from the vantage point of TV, not the weekly short-track feature. As a result, they grow up confusing the artificial with the real.

My nephews enjoyed every one of the 300 laps, especially the ones that involved smoke spewing out of crumpled hulks that had once been spotless Monte Carlos and Tauruses. They made up their own minds on whom to "pull for" and whom to hate, basing their considered opinions on factors and variables of their own choosing.

"I can't stand that guy," said Jake, referring to a driver who shall remain nameless.

"Why not?"

"He's mean, and you can't win no race being mean like him."

"Don't think so, eh?"

"Shoot, no. Them other drivers ain't gonna put up with it."

The kids made their choices largely based on paint jobs. Obviously, Jake was happy that Dale Earnhardt Jr. did well, but he was partial to the black, yellow and white paint job of Tony Stewart. Vince found himself drawn to the blue-and-green No. 37 of Mark Green. Ray found the red, white and blue colors of Robert Pressley most attractive.

The P.A. announcer kept referring to Hickory Motor Speedway as "America's most famous short track," and based on one glorious afternoon with three kids, I guess the description was all right by me.

A PRESS CONFERENCE AT ALCATRAZ? GO FIGURE

june 1997

ALCATRAZ ISLAND, CALIF.— The president of Sears Point Raceway, Steve Page, stood on a balcony next to three stock car racers—Dale Jarrett, Robby Gordon and Rick Mast—and welcomed the attending media to "this most magnificent of all possible press-conference venues."

Magnificent? The drivers were standing in a cell block, surrounded by clanging steel bars and peeling beige paint. Through the windows could be seen the breathtaking skyline of San Francisco. Now that was magnificent.

This was bizarre.

Alcatraz, 34 years after being closed and 29 years after the Indians occupied it, is now a tourist attraction. The beauty of it, from the State of California's viewpoint, has to be the low maintenance. It's a prison, for gosh sakes. The more it rots, the better. Surrounded on all sides by the shark-infested, 50-degree waters of San Francisco Bay, the island is as foreboding as the Bates Motel. As far as the eye can see, Alcatraz Island is the only thing that is ugly.

Only in California can you go to a tourist attraction and be briefed by a male park service ranger with a ponytail that cascades halfway down his back. Can't remember the guy's name but it must have been Lance. Or Troy, perhaps. Definitely not a Clarence.

Bruton Smith and H. A. "Humpy" Wheeler were in attendance. Their publicly traded company, Speedway Motorsports, now owns Sears Point, and with the two great stock car showmen nearby, you had to stay alert. Would a team of commandos parachute in to rescue Jarrett, Gordon and Mast, thus saving (gasp!) Sunday's stock car race? I watched Ranger Lance warily. He seemed a likely candidate to snap and fall into the clutches of evil-doers. Would Clint Eastwood save the day? Sean Connery? Dale Earnhardt?

Ah, the plot was flawed from the start. Any NASCAR fan knows that Earnhardt would never have rescued Jarrett and Mast. They both drive Fords.

Actually, the fire-and-magic was kept to a minimum. These park rangers probably never even heard of Charlotte Motor Speedway and the annual Invasion of the Tri-Oval. They would agree to nothing more flashy than opening the cell doors and having the drivers gape at their mock incarceration as cameras clicked and bulbs flashed.

Bruton walked up behind me to make sure I knew he was offering a $1.8 million purse ("You know, that's unprecedented for a road course.") Then we watched as some West Coast photographer balanced himself with his legs wrapped around the cell-block rail, three stories up, so that he could snap Jarrett from a slightly better angle.

"I want you to look at that idiot," Smith said.

As we left the island and boarded our ferry, they handed out black-and-white striped caps that read:

PROPERTY OF

ALCATRAZ

FEDERAL PENITENTIARY

UNLISTED NUMBER

I put the cap on and waved to Bruton.

"At last," I yelled, "you've got us sportswriters right where you want us."

The tycoon clapped his hands and laughed.

THIS AIN'T THE NASCAR I KNOW
december 1998

NEW YORK, N.Y.— A million stock car racing fans would have killed for a ticket to the annual Winston Cup Awards Banquet Friday night, and for the life of me, I don't know why.

From 6:30 to 11 p.m., I sat at my table in the balcony, thinking I'd just as soon spend the time shoveling out the stables on the family farm.

My favorite thing about the banquet is the menu. When one attends a banquet at the Waldorf-Astoria, a tossed salad is not enough. You get "Mixed Greens with Grilled Asparagus Tips and Sliced Tomatoes served with Champagne Vinaigrette Dressing." Along with the "Grilled Filet of Beef in Morel Mushroom Sauce" are "Oven Roasted Au Gratin Potato" and a "Seasonal Medley of Vegetables," none of which taste like they were cooked more than 30 seconds or with a slab of fatback in sight. Then there is the "Flour-less Cake wrapped in a Dark Chocolate Collar with Light and Milk Chocolate Mousse" and "Raspberry and Vanilla Garnish."

It sure is pretty, but it doesn't taste that great. I wonder what they would have thought had our table ordered out for pizza.

The banquet is relentlessly rehearsed. This year for the first time they had teleprompters installed on the stage so that none of the speakers would say one word that had not been ghost-written (poorly, I might add) and pre-approved. Jeff Burton, to his everloving credit, departed from the prepared text, thus providing one of the evening's few moments of subtle humor.

"They tell you what to say," said Burton later, "but that doesn't mean you have to say it. I speak for myself."

When Burton started to toss in a few quips, he drew some worried looks from the handlers posted nearby.

Each speaker had what was supposed to look like a small chitchat with R. J. Reynolds Tobacco's Cliff Pennell as he arrived at the podium. Have you ever tried to chitchat from a script? It had that "Hord Hardin with the Masters champion" look to it.

Only the last two hours received national television cover-

SCENES ● 191

age, which was wise, but I still don't understand how ESPN can justify televising such an overtly commercial recitation of sponsors at an affair in which nothing is newsworthy. Every winner is known in advance. The champion? Why, Jeff Gordon. The top crew chief? Ray Evernham, of course. Engine builder? Team Hendrick's Charlie Siegars. Sponsor? Yep, DuPont.

Why is it this way? Of course, if there were any surprises, NASCAR would not be able to stomp the life out of the whole evening by scripting every word, every gesture, every symbol.

Between the first segment and the TV portion, the audience was treated to a concert by pop singer Karla Bonoff. How was she chosen? I could see this hush-hush meeting in Daytona Beach.

"Who have we got to choose from?"

"Well, I've got a whole list of artists available. How about Brooks & Dunn?"

"Too country. We want to put across a more uptown image."

"ZZ Top? Springsteen?"

"Nah, too loud and unruly. We've got a lot of Madison Avenue executives in the audience. We don't want to scare anybody."

So after crossing off every artist or group that was the least bit politically incorrect, they probably plucked Karla Bonoff out of some Greenwich Village coffeehouse and dictated that she sing the five blandest songs in her repertoire. That's the way NASCAR operates these days.

If this banquet had a theme, it would be: "New Money in the Shrine of Old Money." You never saw so many radical tuxes. Wearing a large studded button at the top of an uncollared tux is popular among the fashion adventurous. Me, I just ordered the same basic tux I wore last year, which was almost a big mistake because I recently dropped an inch or two in the waist and came dangerously close to losing my pants the first time I walked down a flight of stairs.

Dale Jarrett, who delivered the best speech, wore a midnight-blue shirt that made him look like a priest from the Church of Scientology. Pierre Cardin I'm not, but his getup looked one baby step shy of a Nehru jacket to me.

After a while, I didn't really listen. I just picked out odd little nuggets. Such as John Hendrick, who kept referring to his pit crew as the "Rainbow Wah-yuhs."

Most speeches went like this:

"I'd like to thank Raybestos brakes, who have been a big part of my life ever since I got started in this business. Then there's the fine people at Craftsman Tools and Sears Roebuck. They've stuck with me all the way. I'd like to thank Goodyear, makers of the best tires on earth. And, oh yeah, don't forget the motor-home company that gave me such a good deal. [Looking out into the audience] Mary Ann, what's their name? Winnebago! That's it."

Tirico, despite the fact that he was virtually the only African American in the room who was not refilling water glasses, was masterful. His appointment was another likely made by committee decision of image makers. NASCAR had Tirico pay homage to MRN's Barney Hall, thus reminding everyone in the room that they had shoved Hall out of his traditional emcee's role. He did get a bit overzealous one time, though, when he called the Winston Cup "the biggest story of the 1990s."

What about the fall of communism? The AIDS epidemic? The collapse of economic stability in the Pacific Rim? Our two-term president, the sex fiend? I mean, aren't we getting just a tad pompous in calling a bunch of guys driving 'round and 'round "the biggest story of the 1990s"?

Or is it just me?

BOREDOM IN THE FAST LANE
july 1997

LONG POND, PA.— I know how little patience you fans have for my problems. I know how much you would give to fly back and forth across the country, just for the privilege of watching that most righteous of apple pie-eating, John Wayne-loving, made-this-country-what-it-is-today sports.

Ah, say it with gusto. Beat on your hairy chest and give it that Tarzan touch.

NASCAR! Forever and ever, amen.

Maybe I am spoiled.

That having been said, the Pennsylvania 500—the race that not even corporate America would sponsor—reminded me of the county backgammon tournament. Actually, that may be a disservice to the hard-driving, thrill-a-minute sport of backgammon. OK. The county checkers tournament. Played at a barber shop. With a barber-shop quartet singing the national anthem. A barber-shop quartet from the old *Lawrence Welk Show*. Play-by-play courtesy of Tommy Newsom. Expert commentary from Junior Samples. (Isn't it amazing how many people out there know what B.R. 549 is?)

Perhaps this lingering feeling on my part stems from the fact that I watched a splendid World of Outlaws Sprint-car joust on TNN Saturday night. Or that getting up at 5:30 a.m. is a prerequisite for arriving at Pocono International Raceway before what seems like the entire population of Mainland China arrives at the race track all at the same time. Or that even New York, the City That Never Sleeps, took a catnap during the first 100 laps of the Pennsylvania 500.

You probably don't think the largest city in the country even watches Winston-Cup racing on TV.

Sure it does. George Steinbrenner watches. Woody Allen watches. Mayor Giuliani watches. With that obnoxious kid sitting in his lap. The Governor of Pennsylvania, Tom Ridge, watches. He rides to the track in a helicopter. That way the state won't have to build better highways to the 2.5-mile, triangular track on the frontier.

Everybody watches the Winston Cup. It's the new National Pastime. The sport for millions of Americans whose sole form of exercise is tooling around in their T-Birds (soon to be Tauruses, or is the plural of that Tauri?) and who thus can relate to more active Americans who do it at 180 mph.

Everyone was at the edge of their seats.

Except me. I was working on an upcoming tabloid. And nodding off. And reading my email. And scribbling down the notable fact that Dale Jarrett, a great American, was still in the lead.

Having reviewed some of the more pertinent facts of Sunday's race, it occurs to me that Jarrett earned a check for $104,570 for this figurative little execution of the rest of the world's greatest drivers.

Wonder how much they would have paid him for flicking the switch on an electric chair? With me in it?

ME AND TONY, DOWN AT THE DIRT TRACK
april 1999

MADISON, N.C.— A big lug in a red driver's suit climbed up in the grandstands after wrecking in the Limited-Sportsman feature, and the fans were on him like ants on a picnic basket. Meanwhile, two rows back, the pole winner of the Goody's 500 sat unnoticed.

Tony Stewart visited 311 Motor Speedway ("The Daytona of Dirt") Saturday night, and virtually no one noticed. The Winston Cup rookie had the extraordinary good sense not to show up in the orange-and-white colors of The Home Depot, his sponsor. He came not for a personal appearance, not to pad his earnings, but just as a fan.

This whole adventure occurred from pure, dumb fate. A friend told me he was headed downstairs to interview Stewart. As he left the press box, I said to my friend, "Ask Tony if he wants to join us at 311 Speedway tonight."

I was kidding. My friend laughed. But he attempted to duplicate the joke and actually mentioned it to Stewart.

"Dirt track about 25 miles south of here?" Stewart replied.

"Yeah, I think so."

"I'm going," said Stewart. "Here's my phone number. Call me about 7:30 and I'll meet you guys there."

My friend very nearly needed attention at the infield media center. When he told me, I almost fainted. Winston Cup drivers simply do not do these things anymore. Many of them have never driven on a dirt track, much less gone to one for fun. If they show up, it's not to watch racing. It's to pick up a hefty retainer for signing a few autographs. Only a sportswriter, one like me who doesn't have a life, would actually spend all day at

Martinsville Speedway and half the night at a nearby dirt track. Or so I thought until Tony Stewart, accompanied by two friends, walked through the gate, smiled and said, "Pick us out a seat."

Over the next three hours, Stewart taught me a lot about dirt-track racing. Turns out that Stewart owns a dirt Late Model and in fact won three major events last year. On a Winston Cup off weekend—the next one is in three weeks—it is entirely possible that Stewart will show up unannounced at a short track near you, not because you pay him to be there but because he hears there is a race track and has a hankering to go fast around it.

Stewart is the last of the old-time racers, descended from A. J. Foyt and Junior Johnson. If life were fair, and it weren't so damn hard to win at the Winston Cup level, Stewart would be racing a stock car on dirt Thursday night, a midget on pavement Friday, a rear-engined Indy car on Saturday and a Winston Cup stocker on Sunday. But he has contracts and commitments, so he gets around the best he can. If he can't drive a race car, he watches other people drive them.

As ridiculous as this may sound, Stewart told me, "I race because I love it. If I make a million bucks, I make a million bucks. But that's not the reason I race."

Here's something even more ridiculous: I believe him.

ROUGHIN' IT WITH YOUNG'UNS
april 1997

GAFFNEY, S.C.— Little Vince fell asleep, using a freshly bought T-shirt for a pillow. Jake looked like an Arctic explorer, wrapped in blankets he had bummed from fans nearby. Ray had defected to the row below, lured by the orange glow of a compact heater.

Uncle Monte sat all alone, watching Mike Duvall, aka the Flintstone Flyer, win the 50-lap main event on the dirt of Gaffney's Cherokee Speedway.

It was the latest episode in the annual series known as, "Monte Takes the Kids to the Race Track."

The nephews, of course, know that their uncle writes about racing for a living. He brings them shirts, caps and 1:64 scale miniatures of their favorite drivers. Ray, soon to be 10 years old, is a Jeff Gordon fan. Jake, 8, prefers Dale Earnhardt. Vince, 6, has his allegiances up for sale now that the Little Caesar's pizza man is no longer emblazoned on the Kranefuss-Haas car. The suspicion here is that a few solid performances by Robert Pressley could put Vince solidly in the Cartoon Network camp.

Their uncle thought about taking them to Hickory for the Busch Grand National race. A family meeting was called, however, and the general consensus was that Hickory might be a bit of a stretch, it being two hours from the boys' home in Clinton, South Carolina

On the way up, Fuddrucker's won the straw pole for official supper site of the Dutton Racing League, edging out Spartanburg's Beacon Drive-In. The crucial factor was Jake blurting out, "Hey, let's go to Fuddrucker's. They let you make your own hamburger there . . . and they got that hot cheese stuff!"

The meal cost $23.23. A dozen cookies to take to the track cost $4.71. Track admission was $30 ($15 for one adult, $5 apiece for the kids), and the vehicle needed filling up. A stop at the automatic teller was shortly required, thus allowing for the unavoidable drinks and a few souvenirs. By the time eyes had been cast on the souvenir stand, the budget had reached three figures.

But it was worth it. The main event in the Modified Four division saw a side-by-side finish. The kid who came up two feet short goes by the name of Kyle Davis. Kyle's grandfather, hall-of-fame mechanic Everett "Cotton" Owens, still turns the wrenches on Kyle's No. 5 Plymouth, the No. 4 of brother Brandon and the open-wheel Modified No. 6 piloted by first cousin Ryan Owens.

Earl "Strawberry" Davis won his third straight feature in the popular Thunder and Lightning division. Strawberry's black No. 5 is immaculately prepared, which also comes as no surprise since Strawberry's "day job" used to be working at Bud Moore Engineering. Mechanic by day, racer by night.

Uncle Monte made a vain attempt to explain all of this to the young'uns.

Ray got a quizzical look on his face. "Well, Uncle Monte, why don't Strawberry get him one of them Winston Cup cars like Jeff Gordon?"

"Well, Ray, it takes a lot of money to drive on the big tracks."

"You could give him some, Unc," said Vince. "You got lotsa money, don't ya?"

"Uh, I think Strawberry just likes racing on dirt."

"Likes to go sliding through them turns, huh?" said Jake.

"Yeah, that's it."

"I don't blame him a bit," concluded Jake.

"Cool."

HEAD FOR THE HILLS, MATILDA! HERE COMES NASCAR!
january 1999

DAYTONA BEACH, FLA.— Until last week I really didn't understand just what culture shock stock car racing must be to residents of this area.

Like most of Florida, the Daytona Beach area is overrunning with old people, thousands upon thousands of them. I saw a line the other day in a Super Bowl story about how Southerners all adore Florida because "that's where Yankees come to die."

I didn't think any more about it, however, until I decided to take in an afternoon movie. It was 1 p.m. Naturally I assumed I could walk right up and enjoy the movie in relative anonymity.

Wrong.

When I arrived at the cineplex, the line stretched 30 yards out into the parking lot. What the . . . ?

A large bus had just deposited several hundred senior citizens on the grounds. Fittingly, the bus had a yellow side panel with a single word painted in black letters: Y-A-N-K-E-E.

For *You've Got Mail,* I was voted as the audience's rookie of the year. I'm 40, and there was no one else within 20 years of me in youth. Some of the seniors exchanged glances, wondering if profanity might not be bad for me to hear. In the popcorn line, one lady with pale-blue hair offered me a lollipop.

I politely declined. "Good boy," said another lady nearby. "All this sugar's no good for you."

When the movie ended, in the row behind me, a woman who looked just like Vivian Vance turned to a man who looked like William Frawley and bellowed, "Irving, now THAT was a movie!"

The hurricanes that occasionally come rolling through these parts must be nothing compared to the influx of a couple of hundred thousand race fans. As matter of fact, race cars actually go faster than hurricanes, and both leave a friendly game of shuffleboard frozen motionless at the starting line. This cozy enclave of tortoises all of a sudden has to cope with hares hippity-hopping around. They ought to rename the Pepsi 400 "Night of the Lepus."

These are, however, feisty people. No one who has ever jockeyed for position at Homer's Original Smorgasboard with a couple of oldsters at early-bird special time should underestimate their willpower and resolve. Zorro never brandished a sword with any more skill than one of these old ladies with an umbrella.

If you're going to play golf down here, make sure you've got a full foursome. A friend and I got thrown together with Dick and Jerry over at Pelican Bay South. They were both retirees from IBM. Jerry had a drive that was infinitely better than mine. Hole after hole, he swatted it 175 yards, straight down the middle. Meanwhile, I was hitting it 240, or just far enough and wide enough to hear it splash. Jerry shot 85. I nearly got shot by an old lady who lived in a ranch-style house that bordered the 15th fairway. Dick had a golf swing that looked like Carl Yastrzemski taking a cut at a Sam McDowell fastball. At the top of his backswing, his clubhead made a little dipsy-doodle. If I tried that, I wouldn't come within a foot of the ball.

Dick beat me, too.

It's February, which means that for most of these people, there is no escaping Speedweeks. I mean, what are they going to do, visit their children in Buffalo? Like I said, it's February. So they retreat to their condos and, early in the morning, while the race fans are still hung over, the terrorists among them go out and let the air out of their tires. Serves the young whippersnappers right, carrying on like they do.

I'm not even going to speculate about what Bike Week is like down here. As the commentator said when the *Hindenburg* crashed, "Oh, the humanity!"

TRADE SHOW IN YANKEE LAND
january 1998

FORT WASHINGTON, PA.— To the feisty fans of eastern Pennsylvania, a short-track legend named Kenny Brightbill draws a greater roar than Jeff Gordon's Rainbow Warriors. They shuffle through the Fort Washington Expo Center by the thousands, gladly plunking down $4.75 for a steak-and-cheese hoagie and queueing up for autographs of Sterling Marlin, Chad Little and Steve Park.

A wet snow falls outside the 13th annual Miller Motorsports show, but it neither sticks to the pavement nor dampens the enthusiasm inside. These are hard-core race fans, oval and drag, dirt and pavement, and they have no patience for winter. Already they have stayed up half the night watching rerun marathons of vintage NASCAR races on ESPN2.

Technically, this is a trade show, but it has just as many characteristics of a carnival. Toddlers putter around the floor in a space roped off for tiny yellow motorized carts. Ten yards away, slot cars, decorated like Winston Cup Monte Carlos, jet around a shiny black road course. Winston Cup simulators tumble and rumble on flatbeds wheeled into the arena. Teenagers take the wheel of remote-control racers; others sample computer simulation games. Highlight videos—*The History of Dirt at Williams Grove* and *Modified Madness* among them—are sampled and purchased. Beauty queens have their own booths, where fans can buy posters, purchase 8×10 glossies or have their photographs taken on Polaroids with smiling, scantily clad beauties representing the likes of Miller and Hooters.

At the Miss Motorsports pageant, little is left to the imagination in the swimsuit competition, but it is not enough to be merely beautiful. Some of the contestants have been placed here by Philadelphia modeling agencies, but in behind-the-scenes interviews, imposters are quickly identified.

One of the Rainbow Warriors asks one contestant, who has listed her favorite track as Dover Downs, "What kind of cars race at Dover?"

Her answer ("Er, uh, NASCAR, Indy cars, I think, drags, big-block, uh, modifications, all kinds, really") draws chuckles as judges write twos or threes, instead of eights and nines, on their ballots.

The eventual winner—it is nearly 10 p.m. when the grueling process ends—is a lithe 26-year-old from Sussex, Pennsylvania, named Kimberly Whitehead. Fittingly, she represents an organization known as Dirt Motorsports.

In a question-and-answer session held on the main stage, a Winston Cup reporter struggles to placate the audience with answers to questions like, "Is Rick Hendrick really sick, or is he just trying to get out of serving time?" and, "Are they gonna run restrictor plates at Atlanta?"

Out on the floor, he is asked every 30 or seconds or so, "So who's gonna win the championship this year?" When he shrugs his shoulders and says, "Based on the results of the last three years, I don't see how you can pick anybody but Gordon," his interrogators shrug their shoulders and grudgingly accept it. Some of them reveal their own preferences with comments like, "Yeah, I know what you mean, but last year should've been Martin's, dammit," or, "You know, Earnhardt's bound to get his act back together. He's gotta."

By and large, these are hardworking, fun-loving folk, and if they pay an extra buck-50 to have rum added to their strawberry smoothies, hey, it's cold outside, and they're just trying to knock the chill off, OK?

They've got everything from shock absorbers and temperature gauges to die-cast collectibles and Pennzoil jackets crammed into their shopping bags. They nudge each other and ogle, sizing up everything from the Winston Cup show car to the restored 1959 Indy roadster to the monstrous Sprint car of Ohio's Luke Castle. Mr. Castle, by the way, is all of 12 years old, and he is almost as tall as the bottom of his race car's wing.

Speedweeks at Daytona is only two weeks away, but the Yankee winter is way too long and so is the distance to Florida. A man needs his racing fix, even in the barren cold.

BACK BEFORE THE LUXURY BOXES
february 1998

DAYTONA BEACH, FLA.— Late one night in February 1952, Chris Economaki wanted to thank William H. G. France for allowing Economaki to provide the track announcing for France's stock car race on the old Daytona beach/road course.

So before beginning the drive back to New Jersey, Economaki drove over to Big Bill's modest home at 29 Goodall Street.

Economaki knocked on the door.

"Who is it?" came a gruff voice from inside.

"Chris Economaki."

"Who?" The door cracked ever so slightly.

"Chris Economaki. I just wanted to thank you for letting me announce your race."

Finally, the door opened. "Aw, hell, come on in," said France.

The entire living room floor was covered with money, mostly $1 bills. Big Bill had dumped the take from the race on the floor, and now he and the family were counting it.

"I thought that must have been a million dollars on that floor," Economaki recalled. "Here I was thanking him because he had paid me $150 to announce three days of racing after I had driven 1,500 miles to do it. A hundred and fifty dollars was quite a salary for a public address announcer in those days."

Big Bill France died in 1992, but his creation, the National Association for Stock Car Auto Racing, continues to thrive under France leadership.

Most of the old edifices still stand in downtown Daytona Beach. The old house is still occupied, though the Frances outgrew the modest neighborhood decades ago. Bill France Racing once operated out of a building now occupied by Froggy's Saloon. The Amoco station the old man once owned still stands, though out of business for several years. At the house on 42 South Peninsula Street that was once NASCAR headquarters, a lawyer has moved in and added on a square brick structure at the front.

Even the Streamline Hotel, where NASCAR held its organizational meeting in 1947, is still located across Atlantic Boulevard from the ocean. Only now it's a rundown hostel with a sign on the door that says: SHOWERS $5.

Long before there was a Daytona International Speedway, a 4.2-mile temporary race course was cleared each year. It ran from the southern edge of Daytona Beach all the way down to Ponce Inlet. The course was simple. Stock cars ran wide open up the beach for two miles, jammed on the brakes for a hairpin curve, then jumped on the gas again for two miles of southbound pavement on Highway A-1-A and turned back onto the beach again. Most races were 40 laps.

Curtis Turner was so entertaining, sliding his car through the rutted sand curves at either end, that Economaki remembers using the legendary driver's swashbuckling style to advantage.

"Each car would only come around every four minutes or so, and the fans would start to mill about," Economaki said. "So I'd time Turner, and shortly before he was due to arrive again, I'd take the microphone and yell, 'It's Turner time again!' and the fans would crowd against the barriers to wait for the great man's return."

Is Bill France Jr., the current czar of big league stock car racing, wistful for the good old days? Well, yes and no.

"In some respects, the better things become, the more the barracudas come after you," the NASCAR president said. "Now it seems like you spend half your day making a dollar; then you spend the other half with a bunch of lawyers keeping another bunch of lawyers from taking it away from you."

One can safely attach quite a few zeroes to those France dollars these days.

"I think a lot of it was more fun back then," he said. "The drivers would tell you the same thing. They didn't have the commitments. When they quit running the race, there wasn't anything else for them to do. They'd go out drinking, go out on the town, go eat dinner with their family or their friends or each other and just have a big time. Not a lot of that goes on anymore.

"As I get older, I don't miss the partying as much, I guess."

Big Bill prepared his son for the task of running NASCAR by having him collect money on race morning. Fans used to camp out on the edge of the course the night before, hoping they could avoid paying. Bill Jr. would walk up through the sea oats and palmetto trees with a local deputy in tow, rousting people and telling them either to fork over $4 or vacate the premises.

At the first Southern 500, at Darlington in 1950, Bill Jr. sold snow cones. Demand was so high that he invented a new flavor: plain.

"The race wasn't half over, and all we had left was strawberry and plain," he said.

Now Daytona International Speedway seats over 143,000, and the France family's sprawling empire is headquartered in a new 60,000-square-foot office building. Soon Bill Jr. and his brother Jim will pass it along to Brian France and Lesa France Kennedy.

Asked if he had attempted to instill the old-fashioned values in the next generation, Bill Jr. answered by reaching into his pocket and pulling out a stack of cards.

"We got this from the football coach, Bill Parcells," said France, 64.

On the card was this message:

"Remember in business that the problems you're up against, personnel problems, emotional problems, scheduling problems, or any problems, the reality is NOBODY CARES what you're up against. The sooner you can put those issues out of your mind, the sooner you can direct your focus toward the real issue: Pushing Your Team to Victory."

Now monetary tributes arrive in Daytona Beach from all the NASCAR outposts. In the old days, Big Bill teetered on the edge of financial ruin several times. Once a man arrived unsolicited to hand the old man a $20,000 personal check, enabling NASCAR to meet its payroll. When most of the big-name drivers pulled out of a race at Talladega, Alabama, in 1969, Big Bill ran his race anyway, then rain-checked the house, meaning that everyone there could return to see another race free. That grandiose gesture also stripped away the old man's reserves.

"There's no question that what saved us on several occasions was just good luck, pure luck," Bill Jr. remembered. "My dad used to say that if you take care of your women and kids, the good Lord is going to take care of you."

Big Bill was a visionary, a risk-taker and a showman. When the members of the Professional Drivers Association did not want to race at Talladega, the elder France strapped himself into a race car and ran wide-open around the new track. When he climbed out, the tires were blistered, but he had made his point.

Bill Jr. is more the businessman. He and his brother have made NASCAR and International Speedway Corporation into two of the world's more profitable businesses. What holds the France empire together is a belief in loyalty. Drivers, car owners and track owners alike consider themselves fully accredited members of the extended family.

"Somebody's got to make sure it all works," said Bill Jr., discussing his sport with a small group of writers and broadcasters. "You folks are the drivers, mechanics and pit crew members of your industry. You're the performers because you write the stories that sell the papers. But there's got to be somebody back home tending to the business, making sure everything gets taken care of. It's the same way here."

ONE OF MANY SAD STORIES ON A SUN-DRENCHED DAY

february 1998

DAYTONA BEACH, FLA.— Through the massive glass facing of the Houston Lawing Press Box, the scene at Daytona International Speedway seemed so light, so frivolous, so very Florida.

Jimmy Buffett music wafted through the air on those few moments when talkative public address announcers had nothing to say. Flags waved in the glistening sun. A GM-Goodwrench hot-air balloon hovered far above Lake Lloyd. Planes pulled banners:

STP SEZ: "GOOD LUCK JOHN!" #43

SHRIMP BASKETS $5.99 MAIN ST PIER

PEPSI: DAYTONA SPEEDWAY'S OFFICIAL

SOFT DRINK FOR 40 YEARS

BANK AND BLUES CLUB HOT BLUES TONITE

But on the track, this day was deadly serious. Thanks to a change in the provisional rules, the always treacherous Gatorade Twin 125-mile qualifying races were even more so. Only six positions in the Daytona 500 would come from qualifying speeds; if a driver represented a new team, or one without adequate provisional protection, he was ashen-faced when he climbed into his car.

Sponsors do not understand it when a car, their company's car, fails to make the Daytona 500 starting field. They do not like to pass up national network television exposure. When the simple math is pointed out—56 cars, 43 available slots in the field—CEOs just smile that ruthless smile and say, "Make damn sure you're one of those 43."

So it was under such pressure that Wally Dallenbach Jr. jumped the start of race one. Dallenbach had cleared Dan Pardus and was half a length ahead of Ted Musgrave, no doubt smiling impishly until the word came down from the tower: "Post the 46."

Dallenbach's day may have started out poorly, but his recovery was miraculous. After a stop-and-go penalty, he came close to being lapped before a caution flag on lap 15 allowed him to catch back up. When Dallenbach avoided a subsequent crash (involving Kenny Irwin, Todd Bodine and Dick Trickle), it put him in position to make the 500.

Which he nonetheless did not do.

Jeff Gordon, who had pitted, steamed past Dallenbach on the next restart with about the same brand of aggressiveness that had cost Wally at the start. When Dallenbach futilely tried to block Gordon, it left him out of line in the draft. Within three laps, he was 20th in a race in which he needed 15th.

Incredibly, there was another crash on the final lap. Ken Schrader's Chevy shot up the banking like a rocket in turn one, breaking Johnny Benson's heart and any chance for the new Jack Roush driver to make the field.

In the melee, Dallenbach made one final, desperate stab. He passed brothers Jeff and David Green between turn four and the tri-oval. But it still left Dallenbach 16th and out of the 500 by one position. Gordon, who didn't need a finish to make the field, was 15th.

Just another simple, sad story from the qualifying races. There were perhaps a dozen, all of which will be soon forgotten. For the unfortunate, it's on to Rockingham.

HELPLESS IN THE PRESS BOX
may 1997

CONCORD, N.C.— It is 9:45 p.m., and I have no idea where my story is.

The rain just stopped. The Coca-Cola 600 has been suspended for more than an hour. The track is glistening black with the sheen of moisture. Tow trucks are churning around in a pathetic attempt to make the asphalt dry again. An hour from now, maybe an hour and a half, there may be racing again.

Jerry Gappens, the friendly Charlotte Motor Speedway publicist, has strolled by several times to assure me there is "a window" out there, a magical clearing that will allow this titanic struggle of steel projectiles to be completed.

On the back stretch, the speedway has installed three moving lamps that send powerful beams of light up into the stratosphere. The beams criss-cross back and forth, intersecting and generally lending an otherworldly air to the darkness.

Maybe the lights are Humpy Wheeler's means to summon Batman when Gotham is in disarray. Or maybe Wonderboy when Charlotte Motor Speedway is in need of salvation.

The faithless are streaming through the tunnel over on the back stretch. What do these people have to do tomorrow that would keep them from sitting in the rain for several more hours so that a grand total of six laps can be run?

The cars are all covered on pit road. It occurs to me that car covers are a status symbol in the Winston Cup Series. The top teams all have fancy-schmantzy covers, most hardly distinguishable from the actual paint jobs beneath them. The tailenders have covers just like the ones seen in driveways of the affluent. These teams aren't affluent, so maybe they don't have driveways.

After all these years of making people mad with my stories, finally I have seen print-media hell. It has wood-paneled release bins in the back, is carpeted and the roof is covered with shiny gold panels.

Ever since Robby Gordon, the swine, kidnapped this race about an hour and a half ago, the phones in the back of the press box have been jammed with writers making urgent plans with the operatives in the home office.

Right now, this could all be over. If Gordon had been able to hang on to the rear of his Chevy . . . wait a minute, that sounds suggestive. If Gordon had been able to exercise proper control on the rear wheels in his No. 42 Monte Carlo, there would have been enough time, before the deluge, to get this hootenanny past the required halfway point.

But nnoooooo . . .

Gordon, who had already conspired with Mother Nature to rain out the Indianapolis 500, had to hit the wall. A wall was hit, and thus a yellow flag waved. A yellow flag waved, and thus the cars were slowed. The cars were slowed, and thus the laps took longer to complete. The laps took longer to complete, and thus the rains came. The rains came, and thus the red flag waved. The red flag waved, and thus the press box rose to scream in unison, "We're doomed."

Gordon, upon climbing from his bent Coors can, sounded positively Shakespearean. Methinks he should be playing *Hamlet,* not driving stock cars.

"The rain has stuck with me," saith the Danish prince. "It won't go away."

A colleague confronts WTBS' Ken Squier.

"We'd be out of here now if this race had been run at a decent hour."

Squier elects to let the comment pass. It is for TV's benefit that this race was scheduled at night.

A bit of levity is in order at this point.

"How dare they put the viewing needs of the American public above the needs of a small coterie of influential newspapermen!" another writer says.

Squier breaks up and returns to his booth.

THE LAST HOOLIGAN RACE
october 1998

CONCORD, N.C.— No one watches the hooligan race from the press box. Charlotte Motor Speedway (CMS) does not provide light hors d'oeuvres, although a hamburger can be had, sans condiments, for a mere $4.50. Friday is working-class day at the track.

The fans huddle in just a couple of sections. The race is featured on TV monitors in the infield, but many watch from the roof of the infield media center. Then there are the infield fans, shading their eyes from the sun, flipping burgers on a grill and standing there shirtless and transfixed in their bermuda shorts, watching the stock cars roar by and wearing expressions that seem to say, "There goes the neighborhood."

The 40-lap last-chance race, which determines the final eight starters in the Saturday Busch Grand National (BGN) event, is close to being the last survivor of a dying breed. Over the entire history of NASCAR, more and more starting fields have been determined by qualifying, which basically distances the sport from its roots on the short tracks, where a system of heat races is the preferred means of winnowing out the pretenders.

NASCAR officials have assured anyone within earshot that this is the last year they are going to allow CMS to hold its last-chance races, saying that it costs too much money when the inevitable shenanigans occur.

The last-chance race, usually watched by only a smattering of observers, is exciting. Jason Keller, a BGN veteran at age 28, managed to slither by Bobby Hillin on the last lap to claim the eighth and final transfer slot, enabling Keller to start 43rd on Saturday.

Yes, a lot of sheet metal was crumpled. Desperate men (and women, Patty Moise being in the field) do desperate things.

Meanwhile, a motley collection of scribes, some of them fresh off the golf course, watch from the roof, gazing through binoculars, scribbling notes and waxing sarcastic.

"Uh, oh, look at so-and-so moving up on whatchamacallit," one wizened old wag might exclaim. "There is bound to be trouble there."

And, sure enough . . . sscccrrreeeccchhh . . . boom!

Note how the names have been changed to protect the innocent.

Around and around they go, and where they skid to a stop, nobody knows. Meanwhile, up on the roof of the media center, it looks like the world's only rotary tennis tournament. Instead of the crowd watching the little ball bounce back and forth, though, they turn around in circles, pretty much in a synchronized fashion, the better to watch the prospects dice among the suspects.

Ever seen a kid experiment with dizziness by turning himself in a circle until he loses his balance? The action on the media-center roof is only slightly less wacky. Give everyone a good hit of helium, so they could talk like munchkins, and the scene would be one of perfect disorder, not unlike a Mel Brooks movie.

An added feature of this race, which NASCAR apparently finds barbaric, is the fact that, unlike the late Sportsman Division, no one has ever failed to survive a BGN hooligan Friday.

A kid can't learn how to drive on a superspeedway by driving a DIRT modified on a bull ring. He can't learn how by driving a Dodge Neon in some parking-lot gymkhana, and he sure can't learn by taking his mother's Ford Fairmont to the local drag strip.

If it ain't broke, don't fix it.

Hooligans rule.

LOOSE AMONG THE TEXANS
march 1999

JUSTIN, TEXAS— The tourism slogan of the Great State of Texas is, "Like a Whole Other Country," and the Texans are not bragging.

While I wish North Carolinians had the same pride in their

native state, Texans do err on the side of the pretentious. It is a cracking good place to visit, however.

On Thursday, I ventured deep into the heartland. The capital, Austin, is a good three hours from Dallas-Fort Worth but well worth the trip.

In a state otherwise distinguished by scrubby, bleach-dry conditions, Austin is a vision of beauty, surrounded by wooded hills and humidified by a quaint river.

Texans take pride in their music. With typical bluster, Austin bills itself as "the live-music capital of the world." There is some truth to what they say, however. The area is home to Willie Nelson, Guy Clark, Jerry Jeff Walker, Billy Joe Shaver, Jimmie Dale Gilmore, Robert Earl Keen, the Geezinslaws and Junior Brown.

You may never have heard of any of them, but they don't care. All of them are big in Texas, and Texas is big enough that they can make a comfortable living whether you listen to them or not.

Realizing that this column is not featured in the lifestyles section, I will move on, noting only the existence of a culture indigenous only to this mammoth chunk of real estate. I would have driven to Austin to see Charlie Robison or Shaver and paid $25 to see either. Instead, at a nightspot on Lake Travis known as Carlos & Charlie's, I saw both for free.

Texans do it their way. They've got their own music, their own food, and as ridiculous as their notions may be, they cling to them with stubborn, sometimes even self-destructive, pride.

Travis to Crockett: "I don't care how many Mexicans there are, we ain't gonna give up this dadburn fort."

Crockett to Travis: "Right on."

As an example, witness the Texans' ridiculous insistence on daubing barbecue sauce exclusively on beef. They might marinate the occasional chicken in the stuff, but the misbegotten denizens of the Lone Star State cling to their notion that pork is fit for nothing except sausage.

In the rest of this country, we know the absurdity of this notion. Truly, I pity them.

God love them, they've got their own speedway, a sparkling edifice sitting in the midst of a great plain north of the cattle capital of Fort Worth.

One year the rain came down. One year the water seeped up. They've reconfigured the track and dug up the asphalt and slapped a new coat down. They've paved the parking lots and talked about ingress and egress. They've met with the legislature and commissioned studies and built condominiums and zoned out the rabble. In other words, they've tried every way a Texan knows how to fix this place, and it's still a big old pile of brick-and-mortar boondoggle.

It's just exactly 99 and $^{44}/_{100}$ percent pure Texas. Don't fight it, baby. What they insist on calling a stock car race today isn't going to be anything more than a glorified cattle drive. It's going to take four hours to get in here and five to get out. Pontiacs and Chevys will go sliding through the tri-oval grass, and the fans will commence to 'rasslin' in the infield.

Somehow the republic will survive.

THE WINSTON IS INTENTIONAL MAYHEM
may 1998

CONCORD, N.C.— The Charlotte Motor Speedway (CMS) press box is on the same level as the luxury suites, which means that cynical scribes get the opportunity to mingle with the upwardly mobile in the elevators and hallways.

Encountered on the elevator Saturday was a model family: clean-cut dad wearing plaid shorts, white golf shirt and cap; attractive wife; two cute-as-a-button little girls, one wearing a rainbow on one cheek and a strawberry on the other.

Six hours later, Mommy was screaming for Jeff Gordon, Daddy was questioning Gordon's manhood, and the little girl with the painted cheeks was hammering on younger sis for laughing when Dale Earnhardt spun.

This all-star race has the insidious knack of reducing otherwise normal human beings into screaming, drooling lunatics. That's the plan.

The Winston was concocted by a pair of visionaries, the late T. Wayne Robertson of RJR Nabisco and H. A. "Humpy" Wheeler of CMS. For publicity purposes, this is stock car racing's All-Star Game. For practical purposes, it is an entry-level event in the great fan indoctrination program.

In these parts, stock car racing is a mainstream sport, the very lore of the Carolinas being wrapped up in tales of moonshiners and "revenuers," the Petty Dynasty, "Ol' D.W.," the Wood Brothers Mercury, the "Alabama Gang" and "Fireball."

As hard as it is to believe, however, surveys show that many people, even here, do not love stock car racing. These unimaginative folks are not thunderheads but rather dunderheads. To them, the Daytona 500 is just a bunch of cars going around and around for three hours.

Imagine! There are those who cannot understand the poetry of the draft, the nuances of pit strategy, the righteous rock 'n' roll of a finely tuned engine. No doubt they are the same folks who once dubbed rural electricity wasteful and impractical.

But this race is for them. Pass out the cut-rate tickets. Get them in the gate for no other reason than to see what all the commotion is about. The Winston is simple, short and exciting. Its format has been honed for 13 years. Robertson made the final changes prior to his untimely death in January.

Slowly, they bring the 150,000 or so fans to an emotional boil. The racing starts at 4:30 with the mild excitement of the Legends and Bandolero cars. Then come the Midwestern superspeedway novices of ARCA (the Automobile Racing Club of America). The Winston Open showcases the second tier of the Winston Cup and gives the non-winners a chance to shine. Then the triple-segmented, Saturday-night short-track feature madness of the Winston.

All the while, Wheeler and his operatives have been stirring the pot with fireworks displays, military jets, screaming commentators and gaudy introductions.

By the time the Winston gets under way, some of the good old boys have been pounding Budweisers for, oh, about eight hours. Several years ago, a few of them crept down to the fence in the tri-oval and indelicately doused Earnhardt as his wrecked car sat lodged against the concrete wall. No word on what happened to those wacky adventurers when they returned to their seats. If they were summarily executed in a flagrant act of vigilante justice, no Tarheel jury would convict the killers.

It's a family sport.

THE FAMILY THAT RACES TOGETHER . . .
july 1998

MYRTLE BEACH, S.C.— I write about stock car racing, and I think it doesn't affect them, but it does.

Every year I take them—Ella, Ray, Jake, Vince and Sammy—to the beach for eight days, or the break between first New Hampshire and second Pocono on the Winston Cup schedule. Keeping them in line would be as impossible as understanding the NASCAR rules, so my mother comes along, and one sister for the first few days.

Ella, 14, and Sammy, 1½, are the children of one sister. Ray, 11, Jake, 9, and Vince, 7, live with the other sister, and my mother, across the pasture from my house. The latter group is the product of an on-again, off-again marriage, and they haven't lived with their father in several years. I do what I can to provide a male influence, but it's tough because I'm gone all the time. The annual trip is expensive, but really, it's the least I can do. I need them more than they need me, if the truth be told.

Last Christmas I bought the two older boys a go-kart. From time to time, I take them for a visit to a nearby race shop. The boys' differences can be seen in their choice of drivers. Ray is a Jeff Gordon fan; Jake prefers Dale Earnhardt. Vince, the youngest, voices no particular allegiance, but since he is the introvert, it's difficult to say.

They fight as though each represents his own warring tribe. I remember my own boyhood, when my younger brother and I disagreed on almost everything. But we were close, each understanding that one was precisely what the other was not. We fought in each other's behalf on several occasions, and I worry that these three boys would not. Sometimes I wonder if they even love each other at all.

This year the centerpiece of the whole trip is a visit to NASCAR Speedpark, a sprawling, ambitious complex with tracks and vehicles for virtually every age group.

Every age group, not every body type. For instance, there are no vehicles easy for a 300-pound 40-year-old to climb into. Once I pry myself into the oval-track ride known as "The Competitor" (16-and-older, must be at least 5-foot-4), I discover that the roof will not close above my head. I take off anyway, wrestling the wheel as the roof keeps bumping me in the head. I never have to "lift," in the racer's vernacular, primarily because 300 pounds provides plenty of "downforce." All agree afterward that I have acquitted myself nicely, but I walk around the park with a headache, mildly nauseated, for about the next hour.

Vince requires an apprenticeship in "The Qualifier," essentially a kiddie ride, before he works himself up to "Young Champions," a road-course ride where he can compete with his older brothers. Ray and Jake have both sampled harder rides, but the final hour of the day evolves into a series of showdowns here.

My mother and I watch closely. Jake squeezes by Ray soon after they leave the pits, but two laps later, Ray returns the favor and a tremendous dogfight ensues. Vince, meanwhile, spins out a couple of times, but he seems as fast as his brothers once he gets the hang of it. I watch approvingly as the older brothers, so often prone to bickering and tattling on one other, race cleanly. Ray has a slight advantage when the two of them arrive almost simultaneously behind a slower car. Jake manages to squeeze by cleanly, while Ray gets pinned in the guard rail on the other side. From 100 feet away, I chuckle as I watch Jake laughing uproariously in his car. Well behind now, Ray's eyes burn with the intensity of Junior Johnson as he resumes the chase.

And Ray does catch his brother, though he can never get back around him. As the 15 or so karts are flagged back into the pits, Jake is king of the hill, but in a comical moment, Vince, who had been almost lapped by his siblings, slides past Jake at the pit entrance, pretending he is the winner.

As my heart swells with pride, Ray draws my attention to the nearby scoreboard, where the three best lap times are posted. Ray is the fastest. Jake, who won the race, is second. And little Vince, who had to be coaxed into these peppy little chariots, is third.

Each has something to take home and hang on the imaginary mantel of the mind. But I am happiest of all, because I can see in the looks on my nephews' faces that maybe, just maybe, there is hope, after all.

IT'S A LONG WAY FROM WILKESBORO TO VEGAS
march 1998

LAS VEGAS, NEV.— Willie Nelson sounded the same performing "Me and Paul."

The song was written at a time when Nelson was little more than an itinerant Texas troubadour, writing songs and surviving brushes with the law. Subsequent fame has given a measure of respectability to the old outlaw.

But Willie doesn't play Luchenbach every night, and on this occasion he was playing the Orleans, a Vegas casino. As wonderful as the show was, something had been lost in the translation between the ramshackle dance hall and the sanitary splendor of the casino.

The same could be said of NASCAR.

Or, in other words, I wonder what they're doing in North Wilkesboro this morning.

Is the sport becoming just another Vegas attraction? In the 100,000-plus hotel rooms of the Strip, is this just another option?

"Well, honey, Tom Jones is taking the weekend off. I was looking in the *Review-Journal*. They've got a story on 1-A about this NASCAR thing. What's that all about?"

The Winston Cup Series is not the garden-variety Vegas attraction. Not yet.

The race fans in the stands have streamed down from Utah, Oregon and even Alaska in their trusty Winnebagos and Airstreams. They bought up tickets in a few hours when Las Vegas Motor Speedway secured its date last autumn.

The suspicion persists, however, that this desert oasis is not naturally conducive to the values that have made stock car racing so upwardly mobile a sport. Las Vegas is where empires rise on the broken bones of the working class, not the sweat of honest labor.

Over a midnight serving of steak and eggs at a casino coffee house, I observed a series of broken-down waitresses trundling about. One of them, poor dear, let me get halfway through my meal without bringing the tall glass of milk I had ordered.

I could envision the waitress 30 years ago, probably fresh off a train from Pekin, Illinois, or Medford, Oregon, standing in line for a chorus tryout. Now here she is, old and outdated, living out the rest of her days shoveling hash to surly hillbillies.

Another woman, still in the polished, low-mileage stage of her Vegas career, explained the plight of the young blackjack dealers on the casino floor, many of whom earn $300–500 a night in tips. Most live in shabby, one-bedroom flats on the outskirts of town. Each night after the shift is over, she said, they return to the tables and lose at one end of the table what earlier they had raked in from the other.

Everyone wants to be a big shot out here, but not many of the arrogant millionaires made their fortunes in Vegas. They came here to flaunt it.

Is this true of NASCAR as well, barreling into town like a panhandle tornado and vowing to lay waste to yet another unsuspecting principality?

The boom days don't last forever. The oil gushers ran out in Texas, and eventually the graphs and charts will level out in

Daytona Beach. When that day comes, they better hope they've got a few places left like Martinsville and Darlington to wave their flags.

Las Vegas will turn on them in a heartbeat.

A GLIMPSE AT THE FUTURE
december 1999

Where is this sport heading? In its headlong advance into mainstream status, will the very qualities that made NASCAR so appealing to the masses suddenly become lost in all the money and glitter?

In many respects, they already have, but the corruption of greed and power has not undermined its ever-increasing popularity—yet. The most disquieting sign is the *de facto* philosophy that has governed NASCAR decision makers: what is best for the sport is what makes us the most money. In Daytona Beach, they pay lip service to precaution, but there is no tangible evidence that anyone is lobbying effectively on behalf of the long-time fans.

In the wake of a comprehensive television deal that could bring in nearly $3 billion over the next six seasons, NASCAR has abandoned the companies, principally CBS/TNN and ESPN, that fueled its arrival on the national stage. The response to criticism is typical: the companies had a chance to submit bids. They've been profiting off of us for years. We don't owe them anything. So, now, the prime players in the new game will be NBC and Fox, both of whom avoided any affiliation with stock car racing until, all of a sudden, the rest of the world was marveling at its growth.

NASCAR has blithely ignored its traditions before. The ruling body sat idly by as its ancestral home, North Wilkesboro Speedway, was discarded. While Bob Bahre and Bruton Smith bought up the old track and stole its dates for tracks in New Hampshire and Texas, NASCAR's response was to shrug its shoulders and symbolically say, "Well, times change."

Business as usual, you say? Maybe so. But the National

Football League has never abandoned Green Bay. Wrigley Field still prospers. Muslims do not pray toward Cairo.

What NASCAR has done is stride brazenly toward a monopoly that would make Bill Gates blush. International Speedway Corporation (ISC) and NASCAR are virtually indistinguishable from one another. ISC now owns 10 of the 21 tracks where Winston Cup races are held: California, Darlington, Daytona, Homestead, Michigan, Phoenix, Richmond, Rockingham, Talladega and Watkins Glen. These venues hold 16 of the season's 34 races. Of the 11 other venues, only Smith's Speedway Motorsports Inc. (Atlanta, Bristol, Charlotte, Las Vegas, Sears Point, Texas) stands in uneasy opposition. Indianapolis is aligned with ISC in the construction of a Chicago-area track. The France family owns considerable stock in Martinsville. Pocono, New Hampshire and Dover are reliable ISC allies, and by the time this is published, at least one of these tracks may have been bought by ISC.

In practical terms, the message is clear from Daytona Beach, where both NASCAR and ISC are headquartered. You want a race? Sell us your track. Homestead desperately needed a race, sold out to ISC and Penske, and got one. The Penske consortium wanted to protect its interests, so it was absorbed by ISC. The next tracks scheduled to get races (probably in 2001)? Kansas City and Chicago. Care to guess who owns them? Donald Trump wanted to build in the New York City area? Care to guess what huge speedway corporation he threw in with?

The new tracks, by and large, provide a quality of racing that is inferior to that which transpires on the old tracks. They are luxurious multi-purpose facilities that, like the baseball/football stadia fashionable in the 1960s, do none of purposes well. Almost everyone in the sport is concerned about the deteriorating quality of the spectacle save those responsible for it.

At present, NASCAR and its thinly concealed satellites are vividly preoccupied with the trappings of the sport. Efforts are afoot to handle centrally the marketing and licensing needs of all of the drivers and all of the teams. Theme parks, restaurant chains, interactive gaming, cable-TV channels, pay-per-view programming, clothing, publications and web sites are all being examined intensely. Making it once again possible for race cars to run side-by-side is apparently not.

One can imagine a NASCAR fat cat standing up at a board meeting and exclaiming, "Goodness, people, there's no money in racing! Let's move on!"

NASCAR has developed the paranoia that comes with totalitarian control. Whenever possible, and sometimes by means of tactics that are none too subtle, the sanctioning body has systematically attempted to manipulate coverage of its events. Oh, yes, was it mentioned that it owns its own radio network? That it takes a strong "advisory role" in every aspect of television coverage? That the term "officially licensed" carries with it a vivid connotation of editorial influence in the content of various publications?

Yet, in spite of all the restrictive influences, the sport continues to gallop onward and upward, gleefully counting its money and growing ever hazier about its past. A proposed superspeedway in the New York area, the Trump/ISC joint venture, could seat as many as 200,000 fans.

Meanwhile, in the little hamlets where this sport was born—the Darlingtons, Rockinghams and the Martinsvilles—the clock is ticking. Soon these tracks are likely to have their two annual races scaled back to one. After all, as powerful as NASCAR is, it still cannot adjust the annual calendar so that there are more weekends available for the "multi-faceted" celebrations that also include races.

The clock is also ticking for the embattled, working-class fan, who has been under attack for at least a decade as NASCAR has systematically attempted to price him and his family out of the interactive media stations that once were called grandstands. When this poor, addicted fan hears talk of "premium programming" and infields with "all the conveniences of a small city," he instinctively hears the ch-ching of cash registers.

The scene is familiar every time NASCAR opens one of its gaudy, officially-licensed cafes or go-kart tracks or souvenir stores. Fans press up against the ropes, futilely clutching autograph books and hoping for a wave, a nod, from their heroes. Inside, in the so-called VIP areas, the drivers are ushered in to rub elbows with the beautiful people, almost all of them dressed in hairstyles and clothing that would appear otherworldly in the town square of Darlington, South Carolina, or Martinsville, Virginia. Are they fans? Hell, no. They're agents, and brokers, and people who claim there actually is a profession called "entrepreneur." They're the courtesans of NASCAR's decadent *nouveau riche*.

In some ways, the best thing that could possibly happen is the end of this ungodly boom. If there are empty seats in the glittering edifices, if the television ratings start to wane, then perhaps someone in Daytona Beach will wake up. Otherwise the goose that laid this golden egg will be found flattened out on Interstate 95, the victim of a sneak attack by a Rolls-Royce.

• **index**

Aaron, Henry, 185
ACC, 151
Alabama Gang, 5, 26, 27, 214
Alcatraz, press conference, at, 189–90
Allen, Irwin, 183
Allen, Woody, 194
Allison, Bobby, 27, 69, 97, 136
  on racing, 5–7
Allison, Davey, 107, 169
  Alabama Gang, 27
  death, 28
  rookie win, 37
Allison, Donnie, 27, 72
AMA races, 103
American Speed Association, 62, 115
American Zoom (Golenbock), 66
Anaheim Angels, 120, 121
Anderson, John, 110
Andretti, Aldo, 156
Andretti, John, 18, 44, 155–57, 156, 157, 159, 163, 165
  1997 Daytona 500 win, 133
Andretti, Mario, 155, 156, 184
Andretti, Michael, 155
Andrews, Paul, 54
Anne Arundel County Track, (planned), 141
Apperson, Bob, 16

Area Auto Racing News, 109
Arness, James, 182
Atlanta Motor Speedway, 31, 82, 127, 129, 130, 171, 201, 212, 220
  1997 qualifiers, 137
  Hampton location of, 91–92
  races at, 83, 84, 162
Automobile Racing Club of America (ARCA), 62, 65, 214
  1979, winner K. Petty, 60–61
  1998, winner A. Petty, 60, 61
  SuperCar Series, 24

Bahre, Bob, 108, 130
Baker, Buck, 27, 97
Baker, Buddy, 30, 42, 58
Baldwin, Jack, 157
Balmer's Box, 97
Balmer, Earl, 97
Barfield, Ron, 24
BASCAR, 158
baseball, 112, 166
  BASCAR, 167–69
  uniform patches, sponsor, 167
Baseball Association of Stock Car Auto Racing (BASCAR), 158
Beasts, Bob East cars, 79

Beauchamp, Johnny, 67–68
Bell, Thomas Hunter, Jr., 106
Belle, Albert, 122
Belliard, Rafael, 159
Benson, Johnny, 154, 207
Bernard, Crystal, 143
Beverley, Tim, 10
Bickford, John, 30, 37
   on the father-son relation-
      ship, 7–9
Bill France Racing, 203
Bird, Steve, 152
Blanks, Larvell "Sugar Bear,"
   185
Blockbuster Pavilion, 157
Bodine, Barry, 11, 152
Bodine, Brett, 10, 70
Bodine, Geoff, 70, 133, 152, 179
   business realignment, 54–55
   on being driver–car owner,
      9–11
Bodine, Todd, 137, 207
Bonnett, Neil, 28, 110, 136,
   148–49
   Alabama Gang, 27
   broadcaster, 169–70
   death of, 145, 171
   injured at Darlington, 169
Bonoff, Karla, 192
Bosch Platinum Spark Plugs,
   169
Boston Globe, The, 108, 109
Bradberry, Gary, 77
Bradley, Chris, 62
Brickyard 400
   1994, 136
   1997, 101–3
Bridges, Jeff, 41
Brightbill, Kenny, 200
Bristol Motor Speedway, 31, 32,
   83, 130, 143, 161–62, 164,
   220
   description of, 93–94
Brown, Junior, 212

Browning, Chris, 110
Bryan, Jimmy, 143
Bud at the Glen, The, 54, 118,
   119
Bud Moore Engineering, 198
Budweiser, 20, 109, 123, 214
Burton, Jeff, 20, 73, 133, 163,
   191
Burton, Ward, 12–13, 107, 134,
   135, 164–65, 168
Busch Grand National (BGN),
   24, 33, 51, 57, 63, 75, 81,
   87,121, 141, 174, 186, 187,
   197, 210
   car design, in, 115–16
   difficulty level, 37
Busch Grand National division,
   19, 174
Busch North Series, 109
Byron, Red, 27

Cabarrus County motorsport
   symposium, 89
California 500
   1998, 120
   1998 qualifiers, 94–95
California Speedway, 107, 128–
   29, 131, 220
CanAm, 143
Cannon, Melanie, 56
CART races, 31, 103, 148, 155,
   157
Carter, Gary, 185
Carter, Travis, 76, 148
Cash, Johnny, 76
Castle, Luke, 202
CBS, 219
Charlotte Motor Speedway, 8,
   11, 17, 59, 60, 61, 62, 94,
   124,128, 130, 152, 175,
   190, 208–9, 210, 220
   1998 Winston 500, 213–15
   BASCAR announced, 157–59
   local symposium, 88–90

now Lowe's Motor Speedway,
171, 175
R. Phillips, 171–73
Tom Cotter, 56
Charlotte, NC, Stock Car Racing Capital, 151–53
Cherokee Speedway, 58, 187, 196
Chicago races in (planned track), 164, 220
Childress, Richard, 81, 147
Christian, Sara, 142
Chrysler Corporation, 62
cigarettes, brand loyalty, 123
Citgo, 151
Clark, Ed, 127
Clark, Guy, 212
Clemens, Roger, 159
Cliburn, Van, 116
Clinton, Hillary, 122–23
Clinton, William, 177
CMT 300, 38
1998, 159–61
Cobb, Ty, 65
Coca-Cola, 151
Coca-Cola 600, 153
1994, 25
1997, 30, 208–9
Columbia, SC, dirt track, 63
Concord Monitor, 14
Cook, Jerry, 26
Cope, Derrike, 152, 182
Cotter, Tom, 56, 57
Craftsman Tools, 193
Craftsman Truck Series, 24, 38, 57, 141
1995, series beginning, 153
difficulty level, 37
track locations, 152–53
truck design, in, 115–16
Craven, Ricky, 35, 36, 109, 152
profile, 13–15

Dallenbach, Wally, Jr., 155, 156
1998 Gatorade Twin, 206–7

Dallenbach, Wally, Sr., 155
Darlington Raceway, 128, 131, 162, 220
1990 injury, N. Bonnett, 169
description of, 96–97
greatest challenge, 97
NASCAR, and, 221
pre-race golf outing, 67
Davis, Bill, 135, 162
Davis, Brandon, 58, 197
Davis, Earl "Strawberry," 197
Davis, Kyle, 58, 197
Daytona 500, 24, 86, 156, 214
1959, 65, 67
1967, 59, 184
1976, 46
1994, 171
1997, 29, 133
1998, 146, 147
1999, 74, 75
50th anniversary, 12
qualifier race, 206–7
Daytona Beach area
1998 forest fires, 97–99
early temporary track, 203
race time upheaval, 198–200
Daytona International Speedway, 42, 44, 59, 60, 74, 75, 76, 114, 131, 136, 145, 151, 156, 183, 192, 207, 220
1998 qualifier races, 206–7
death of N. Bonnett, 145
headquarters, 204
Daytona of Dirt, The, 195
Daytona USA exhibit, 29
Dennis, Bill, 16
DeVilbiss 400, 77
DieHard 500, 1993, 170
DiGard, 39
Dirt Motorsports, 201
Dirt tracks, T. Stewart driver, 195–96
DMF Communications, 57, 212

Donlavey, Junie, 137
    profile, 15–17
Dover Downs International
    Speedway, 11, 83, 134, 141,
    144, 201, 220
    1981 winner, Jody Ridley, 16
    1997 Ricky Rudd win, 72
    24 hours of Dover, 99
    description, of, 99–101
dragsters, 156
Drew, Brice, 137–38
Drew, Homer, 137–38
drivers
    private lives, of, 138–39
    track versatility, 196
DuHamel, Yvon, 16
DuPont, 160, 192
Durocher, Leo, 16
Dutton, Monte
    family days, Speedpark, 215–
    17
    family race days, 186–89,
    196–98
    on being a sportswriter, 173,
    177–80
    personal favorites, 184–85
Duvall, Mike, 196
dynamometers, 181

Earnhardt, Dale "The Intimi-
    dator," 11, 13, 15, 19, 21,
    23, 39, 47, 48, 49, 53, 54,
    57, 58, 68, 69, 76, 82, 85,
    97, 102, 109,114, 121, 124,
    125, 129, 136, 149, 150,
    151, 157, 162, 169,170,
    171, 190, 201
    advice to Gordon, J., 6–7
    career length, of, 146–47
    Elliott rivalry, 23
    injury at Talladega, 146–47
    NASCAR 50th anniversary
    win, 12, 136
    profile, 17–18
    scrape with Gordon, 29

Earnhardt, Dale, Jr. "Little E,"
    38, 188, 189, 197, 213, 215
    Great White Hope, 20, 125
    profile, 19–22
    Winston Cup debut, 19
Earnhardt, Ralph, 17, 26, 58–
    59
Earnhardt, Teresa, 57
East, Bob, "Beasts," 79
EasyCare 100, 61
Economaki, Chris, 202, 203
Edison Park, race fans at, 120
Elliott, Bill, 55, 57, 97, 107, 114,
    144, 150, 157, 180
    profile, 22–24
Elliott, Ernie, 22
Elliott, George, 144
Erskine College, 58
ESPN, 125, 192, 200, 219
Esquire, 41
Estes, Wayne, 93
Eury, Tony, 21
Evans, Andy, 10
Evans, Richie, 26
Evernham, Ray, 44, 126, 133,
    152, 160, 161, 192
    Gordon's crew chief, 8
    profile, 24–26

fans
    at the Winston, 213–15
    being priced out, 221–22
    brand loyalty, 122–23
    driver loyalty, 124
    ruggedness of, 126–27
    stereotype of, 123–24
Fans Against Gordon (F.A.G.),
    53
Farmer, Charles Lawrence
    "Red," profile, 26–28
Fast as White Lightning
    (Chapin), 66
Featherlite Modified Tour, 109
    1998, 104–5
Fellows, Ron, 157

Felton, Gene, 16
Fittipaldi, Emerson, 181
Flintstone Flyer, 196
Flock, Frances, 85
Flock, Tim, 66
  fundraiser for, 84–85
Ford Motor Company, 44, 55,
  87
  acclaims Gordon's car, 132,
    133
  Kenny Irwin vehicle, 38
  motorsports division, 42–43,
    54
  motorsports press confer-
    ence, 180–81
Ford Special Vehicle Opera-
  tions, 54
Formula One racing, 44
  point system, 149
Fort Washington Expo Center,
  200
Fox television, 219
Foyt, A. J., 27–28, 147, 155, 196
France family, 136
  (NASCAR), 65
  drivers, extended family,
    205–6
  generations of, 204
  racetrack holdings, 220
France, Brian, 131, 140, 158,
  204
France, Jim, 131, 204
France, Lesa (Kennedy), 204
France, William C., Jr., 59, 104,
  140, 158, 159, 204, 205,
  206
  NASCAR president, 140
  rivalry with Smith, 130–32
France, William H. G., Sr., 140,
  144
  1992 death, 203
  early days, legacy, 202–6
Frank, Larry, 97
Frawley, William, 199
Fuller, Jeff, 152

Galaxy 300, 1998, 186
gambling, Las Vegas, 176–77
Gammons, Peter, 108
Gant, Harry, 97, 146
Gantt, Darin, 184
Gappens, Jerry, 208
Gaston Gazette, 174
Gateway Race Track, St. Louis,
  141
Gatorade Twin, qualifier race,
  206–7
General Motors Corporation,
  87, 124
George, Tony, 131
Gibbs, Joe, 140, 144
Gilmore, Jimmie Dale, 212
Glidden, Bill, 55
Glidden, Bob, 54–55
Golenbock, Peter, American
  Zoom, 66
Goodrich, Bill, 180
Goody's 500, 32
  1997, 80, 81
  1999, 195
Goody's Headache Powders,
  169
Goodyear, 167, 193
Gordon, Cecil, 50
Gordon, Jeff, 11, 15, 17–18, 19,
  20, 21, 31, 33, 38, 39, 44–
  45, 47, 52, 54, 69, 79, 80,
  82, 97, 102, 107, 109, 121,
  124, 125, 134, 151, 157,
  163, 166, 169, 180, 182,
  188, 192, 197, 198, 201,
  207, 213, 215
  1994 Brickyard 400 win, 136
  1998 Southern 500 win, 73–
    74
  1998 winning season, 133
  early career, 7–9
  fan dislike, 125–26
  Fans Against Gordon
    (F.A.G.), 53
  from Vallejo, Calif., 13

no rookie wins, 37
points win, 149–50
profile, 29–30
racing politics, 131–33
Rainbow Warriors, 200
runs out of gas, 24–25
tires confiscated, 159–61
Gordon, Robby, 156–57, 182, 189, 190, 209
first year races, 74
profile, 30–32
Grand National, 58, 59
later Winston Cup Series, 142
Petty legacy, 66
Grand Prix, 143
Great White Hope
See Earnhardt, Dale, Jr. "Little E"
Green, Jeff, 74
Green, Mark, 189
Greenberg, Hank, 142
Greenville-Pickens Speedway, 141–42
Greenwood, Lee, 159
Griffey, Ken Jr., 159

Haas, Carl, 44, 197
Hall, Barney, 193
Hamilton, Bobby, 134, 135
profile, 32–34
Hamilton, Bobby, Jr., 33, 62
Hamilton, Steve, 124
Helton, Mike, 140, 141, 144
Hendrick Motorsports, 48, 81, 89, 126, 192
experimental chassis denial, 133
Hendrick, John, 143–44, 193
Hendrick, Richard, 10, 22, 71, 73, 151, 201
Hickory Motor Speedway, 186, 187, 189, 197
Hiller, John, 185
Hillin, Bobby, 210

History of Dirt at Williams Grove, The (film), 201
Holman-Moody, 5
Homestead Motorsports Complex, 131, 220
Hoosier Dome midget race, 79
Hornaday, Ron, 152
Horton, Jimmy, 170
Hulme, Denis, 143
Hunter, Catfish, 185
Hyde, Harry, 89, 143

Indianapolis 500, 16, 31, 79
1997, 30, 209
Indianapolis Motor Speedway, 131, 181
description of, 101–3
Indianapolis Star, 103
Indy cars, 44, 128–29, 156, 202
drivers' success in NASCAR, 155
popularity in Calif., 152
spectator fatality, 173
Indy Racing League, 103, 156
1997 championship, 79
injuries, racing with, 35–36
International Motorsports Hall of Fame, 143
1998 inductees, 141–42
Lee Petty's induction, 64
International Race of Champions (IROC), 48, 50, 51
International Speedway Corporation (ISC), 130–32, 205
racetrack holdings, 220
Interstate Batteries, 140
Interstate Batteries 500, 1997, 116–17
Intimidator, The
See Earnhardt, Dale
Irvan, Ernie, 16, 29, 81, 152
from Salinas, Calif., 13
injury at Texas Motor Speedway, 117

profile, 34–36
welcomes Stewart, 80
Irwin, Kenny, 56, 79, 183, 207
    profile, 37–38
Isaac, Bobby, 170

Jackson, Joe "Shoeless Joe," 142
Jackson, Richard, 56
Jarrett, Dale, 20, 29, 43, 76, 81,
        82, 114, 133, 159, 163, 165,
        166, 189, 190, 192
    1997 Pennsylvania 500, 194–
        95
    1998 Pontiac Excitement,
        153–55
    Dale Carnegie graduate, 39
    media relations, 39
    profile, 39–41
    win at Michigan Speedway,
        107
Jarrett, Ned "Gentleman," 40,
        184
    win against L. Petty, 63
Jerry Nadeau Fan Club, 57
Jerry Springer Show, 177
Jiffy Lube 300, 12
    1998, 13
Joe Weatherly Stock Car
        Museum, 96
Johncock, Gordon, 143
Johnson, Buck, 78
Johnson, Jack, 125
Johnson, Jimmy, 133
Johnson, Robert Glenn, Jr.
        "Junior," 5, 10, 23, 58, 68,
        76, 106, 132, 196, 216
    moonshine runner, 41
    profile, 41–42
Justice, Charlie "Choo Choo,"
        106

Kahn's Weiners, 167
Kansas Speedway (planned),
        130, 131, 220

Keen, Robert Earl, 212
Keller, Jason, 210
Kendall, Tom, 157
Kennedy, Lesa France, 204
Kenseth, Matt, 144
Kim Chapin, Fast as White
        Lightning, 66
Kinser, Steve, 156
Kranefuss, Michael, 53, 54, 164,
        168, 197
    profile, 42–45
Kulwicki, Alan, 10, 70, 87, 136,
        143, 150
    1992 Winston Cup Cham-
        pion, 9
Kulwicki, Gerald, 9

Labonte, Bobby, 47, 56, 76, 134,
        135, 136, 140, 157
    profile, 45–46
Labonte, Terry "The Iceman,"
        22, 46–49, 76, 129, 149,
        150, 151, 157, 163, 165
    1996 Winston Cup champion-
        ship, 136
    1998 Pontiac Excitement,
        153–55
    hired by Jr. Johnson, 42
    points at Talladega, 18
LaJoie, Randy, 152
Lake Ward, 12
Landry, Tom, 117
Las Vegas 400, 1998, 128–29
Las Vegas Motor Speedway,
        140, 162, 164, 218, 220
    description of, 103–4
    driver favorite, 128–29
Las Vegas, area description,
        217–19
Last American Hero, The
        (film), 41
last-chance race, 210–11
Leader Bonus program, 134
Leno, Jay, 29

Letterman, David, 29, 60
Lewinsky, Monica, 177
Little E
  See Earnhardt, Dale, Jr. "Little E"
Little, Chad, 152, 184, 200
Loomis, Robbie, 163
LoPatin, Larry
  builder, Michigan Speedway, 106
  builder, Texas World Speedway, 107
Lorenzen, Fred "Fearless Freddie," 96, 184
Lowe's Motor Speedway, 171, 175, 179
Lund, DeWayne "Tiny," 76, 170

Manchester Union Leader, 14
Marcis Auto Racing, 49
Marcis, Dave, 10, 11, 146, 149
  profile, 49–50
Maready, Laura, 183
Marino, Dan, 55, 57
Marlin, Coo Coo, 121
Marlin, Sterling, 74, 113, 166, 200
Martin, Gerald, 146, 147
Martin, Mark, 20, 72, 84, 107, 133, 151, 157, 159, 161, 163, 201
  1998 Las Vegas 400, 129
  profile, 50–52
  tires confiscated, 160
  Winston Select 500, 18
Martin, Matthew Clyde, 51
Martin, Strother, 143
Martinsville Speedway, 32–33, 34, 82, 131, 134, 162, 164, 220
  description of, 104–6
  NASCAR, and, 221
Martocci, Filbert, 81
Mast, Rick, 189, 190
Matte, Tom, 185

Mattei, Jim, 10, 87
Mayfield, Jeremy, 43, 168
  1998 Pocono win, 133
  profile, 52–54
McCall, Buz, 157
McClure, Larry, 32
McDowell, Sam, 199
McLaughlin, 152
McReynolds, Larry, 147
Melton, Ray, 96–97
Michigan Speedway, 128–29, 131, 220
  description of, 106–8
Midget, USAC title, 79
Miller Lite 400, 1997, 107
Miller Motorsports show, 200, 202
Miller, Butch, 153
Miss Motorsports, selection of, 201
Mitty, Walter, 57
Mobley, Ethel Flock, 142
Modified Four, 197
Modified Madness (film), 201
Modified races, 26, 58
Moise, Patty, 210
Moore, Bud, 5, 10
Moore, Dudley, 157
Moretti, Gianpiero, 183
Morgan-McClure Racing, 32
Moroso, Rob, 137
Morse, Lee, profile, 54–55
Most Popular Driver, NASCAR, 23
Motley, Jeff, 166
Motor Racing Network, 118, 129, 193
Motor Racing Outreach, 176
Mountain Dew Southern 500, 1994, 22
Mull, Edward, 124
Mull, Regina, 124

Nadeau, Jerry, profile, 55–57
names, spelled the same, 182–83

NAPA 500, 1997, 91–92
NASCAR, 67, 144, 167, 220
  1998 car owner wins, 144
  50th anniversary celebration,
    65, 82, 136
  commercialism of, 167–69,
    221–22
  early days of, 204–5
  few winners, 133–34
  first organizational meeting,
    203
  first time at Indianapolis,
    102
  future of, 140–41, 218–19,
    219–22
  legitimacy of, 135–38
  point standings, 35–36
  politics of, 131–33, 166
  previously Strictly Stock Divi-
    sion, 142
  qualifying races, increasing,
    210
  relationship with ISC, 130,
    220
  relationship, Speedway
    Motorsports, 130–31, 220
  short track racing, 161–65
  television coverage, 90, 219,
    221
  use of retrotechnology, 115
  Winston Cup Awards, 1998,
    191–93
NASCAR Cafe, 39, 40
NASCAR Speedpark, 216
National Basketball Association, The, 92
National Football League, 57,
  166, 219–20
National Hot Rod Association,
  Pro Stock division, 55
National Motorsports Press
  Association, 68
  interview etiquette, 21
National Sportsman title, Red
  Farmer, 26

NBC, 219
Nelson, Gary, 144, 155, 159
Nelson, Willie, 212, 217
Nemechek, Joe, 10, 31, 74
New Hampshire International
  Speedway, 220
  description of, 108–10
  races at, 12, 159–61
New York Times, 167
Newman, Paul, 143
Newsom, Tommy, 194
Newton, Wayne, 129
NHRA races, 103
Nichels, Bill, 59
North Carolina Speedway
  (Rockingham), 48, 75, 82,
  83, 110–11, 130, 162, 207,
  220
  description of, 110–11
  NASCAR, and, 221
North Wilkesboro, 138
  replaced by Texas Motor
    Speedway, 117

Opel Cup, 56
Owens, Everett "Cotton," 197
  profile, 58–60
Owens, Ryan, 58, 197

Page, Steve, 189
Panch, Marvin, 58
Parcells, Bill, 205
Park, Steve, 56, 57, 148, 200
Parsons, Phil, 145
Pearson, David, 58, 59–60, 68,
  69, 97, 109, 160, 182
  crash with Petty, 46–47
Pearson, Larry, 174, 175, 182
Pennell, Cliff, 191
Pennsylvania 500, 149
  1997, 194
Pennzoil, 123
Penske Motorsports, 130, 131
Penske, Roger, 5–6, 10, 43, 44,
  71, 107–8

Pepsi, 151, 188
Pepsi 400
  1998, 34, 97
  1999, 199
Pepsi Southern 500
  See Southern 500
Perrine, Valerie, 41
Perry, Steve, 16
Petty dynasty, 214
Petty Enterprises, 42
Petty, Adam, 64, 65
  on his father, Kyle, 61
  profile, 60–63
  tragic accident, C. Bailey, 62
Petty, Kyle, 10, 29, 60, 61, 64,
    74, 133, 182
Petty, Lee, 60, 168, 182
  1961 injury, 65
  profile, 63–68
Petty, Maurice, 65
Petty, Richard, 50, 60, 61, 63–
    64, 65, 66, 67, 135, 169,
    182
  200th win, 12, 136
  crash with Pearson, 46–47
  inducted to Stock Car Hall
    of Fame, 68
  profile, 68–70
  spectator fatality, 62
Pettyjohn, Ed, 16
Phillips, Russell, 171–73
Phoenix International Raceway,
    32, 83, 103, 131, 220
Pocono 500, 1998, 43
Pocono Raceway, 43, 52, 131,
    194, 220
points
  applied in BASCAR, 159
  Formula One, 150
  Leader Bonus program, 134
  NASCAR, 149–50
Pontiac Excitement 400, 1998,
    153–55
Porter, John, 10

Press Box Television (PBTV),
    95
Pressley, Robert, 137, 185, 197
Pro Sports Car races, 103
Professional Drivers Associa-
    tion, 205
Prost, Alain, 143
Pruett, Scott, 157
Pure Oil Company, 67

Quaker State Oil, 151

R. J. Reynolds Tobacco, 64, 123,
    191, 214
racetracks, new, inferior, 220
Rahal, Bobby, 148, 183
Ralph Earnhardt Grandstand,
    188
RC Cola, 169
Reagan, Ronald, 136, 169
Reno, Marc, 152
restrictor plates, 113–14, 136,
    201
Review-Journal, 218
Richard Childress Racing, 50,
    125, 170, 171
Richard Petty Driving Experi-
    ence, 126
Richardson, Kristi, 110
Richmond Fairgrounds, 59
Richmond International Race-
    way, 16, 82, 141, 146, 153–
    55, 162, 163, 165, 166–67,
    220
Richmond, Tim, 36, 42, 136
Ridge, Tom, 194
Ridley, Jody, 16
Rio, The, 104
rivalries
  Earnhardt Jr.–Gordon, 20,
    21, 124
  Earnhardt Sr.–Elliott, 23
  fans (Ford *vs.* Chevy), 124
  France-Smith, 130

Robert Yates Racing, 49
Roberts, Edward Glenn, Jr.
    "Fireball," 27, 58, 68, 97,
    214
    death, 169
Robertson, T. Wayne, 214
Robinson, Al, 101
Robinson, Jackie, 142
Robison, Charlie, 212
Rolex 24, 156
    1998, 183
Rookie of the Year
    1997, Mike Skinner, 74–75
    1998, Kenny Irwin, 79
Roper, Rom, 125
Rosberg, Keke, 182
Rose, Pete, 142, 183
Rossi, Mario, Dodge Daytona, 5
Roush, Jack, 10, 11, 71, 73, 81,
    129, 144, 160, 161, 181,
    207
Rudd, Ricky, 10, 16, 133, 134,
    135, 151
    1997 Brickyard 400 win, 102
    on racing, 70–72
    Winston Cup debut, 72
Ruffner, Bo, 180
Rutherford, Johnny, 16
Ryan, Bob, 108

Sabates, Felix, 10, 30, 31, 32
    "Gordo effect," 73–74
    fired Kenny Wallace, 81
Samples, Junior, 194
Sanders, Deion, 112
SCCA races, 103, 157
Schipp, David, 103
Schrader, Ken, 16, 141, 145,
    154, 180, 207
Scott, Wendell, 141–42, 143
Sears Point Raceway, 130, 189,
    220
    1998, J. Burton crash, 165–66
    area description, 111–13

Sears Roebuck, 193
Seifert, Bill, 50
Selig, Bud, 157, 158–59
Shaughnessy, Dan, 108
Shaver, Billy Joe, 212
Shepherd, Morgan, 11, 149
Sheppard, Steve, 34
short tracks
    need for, 161–65
    qualifying heats, 210
    small town market, 162
Shuck, Jeff, 126
Shuman, Buddy, 27
Siegars, Charlie, 192
Silver Crown, USAC title, 79
simulators, Winston Cup, 200
Skinner, Mike, 114, 152
    profile, 74–75
Smith, Bruton, 93–94, 127, 129,
    140, 183, 190
    racetrack acquisitions, 130–32
Smith, Larry, 170
Smith, Louise, 141–42, 143
Smith, Stanley, 170
soaking, banned practice, 160–
    61
Southern 500
    1950, first, 15–16, 204
    1962, 97
    1994 Mountain Dew, 22
    1998, 68, 73
    Darlington Raceway at, 6,
    96
    multiple winners, 97
souvenir sales, 125
    Charlotte, area, 151
    importance of, to drivers, 80
Spassky, Boris, 115
Speed, Lake, 145
Speedway Motorsports, Inc.,
    190
    public relations, 127
    racetrack holdings, 130–32,
    220

Speedway Scene, 109
Speedweeks, 200, 202
Spencer, Jimmy "Mr. Excite-
    ment," 42, 107, 114, 123
    profile, 75–77
sponsors, 6–7, 122–23, 151,
    167–69
    automotive manufactures, 6
    consumer, 6
    driver pressure, 206–7
    small markets, and, 162
Sports Car Club of America,
    Trans-Am race, 156
Sportsman Division (BGN),
    171, 173, 211
Sprint cars, 156
Sprint, USAC title, 79
Sprite, 188
Squier, Ken, 209
St. Louis Speedway, 131
Standridge, Billy
    profile, 77–78
Steinbrenner, George, 194
stereotypes
    drivers, of, 138–39
    fans, of, 123–24
Stewart, Tony, 75, 189, 195
    on why he races, 196
    profile, 79–80
Stock Car Cafe, 40
Stock Car Hall of Fame, Rich-
    ard Petty induction, 68
stock car design, 144–45, 180–
    81
    aerodynamics, 115
Stowers, Jared, 125
Streamline Hotel, 203
Sullivan, Tim, 160

Talladega Motor Speedway, 18,
    31, 32, 34, 35, 52, 63, 75,
    76, 78, 86, 131, 136, 147,
    162, 220

1998, Winston 500, 113–16
Bill Elliott, 23
injuries, 146–47, 170
politics at, 77
rain-checked by France, 205
Talladega Short Track, 28
teams, automotive, attitudes
    between, 132
Texaco Oil, 67
Texas Motor Speedway, 94, 129,
    130, 140, 162, 179, 183, 220
    1997, Winston Cup first,
    116–17
    renovation of, 213
Texas World Speedway, builder,
    LoPatin, 107
Texas, Austin area description,
    211–13
Thomas, Dave, 184
Thomas, Frank, 121
Thomas, Herb, 66, 85, 97
Thunder and Lightning, 197
Tide detergent, 122, 151
tires
    "soaking," 160–61
    confiscated, 159–61
    nitrogen filled, 161
Tirico, Mike (awards speech),
    193
TNN, 194, 219
Tracy, Paul, 183
Trans-Am race (SCCA), 143,
    156, 157
Trickle, Dick, 137, 149, 207
Triple-A minor leagues, 63
Trump, Donald, 220
Trump-ISC, New York track,
    220, 221
Turner, Curtis, 28, 42, 53, 66,
    68, 76, 96, 203–4

UAW-GM Quality 500, 185
United Auto Workers, 124

United States Auto Club
(USAC), 38, 79
racing titles, 79
University of Tennessee, hang-
out, NASCAR Cafe, 40
Unser, Al, 183
Unser, Al Jr., 183
Unser, Bobby, 183

Walker, Jerry Jeff, 212
Wallace, Kenny, 154, 162
media relations, 80
profile, 80–82
qualifying time, Martinsville,
82
Wallace, Mike, 62, 81, 82
Wallace, Russell William
"Rusty," 81, 82, 150, 157
profile, 82–84
wins, 1997, 82, 83
Waltrip, Darrell, 10, 11, 35, 56,
69, 70, 102, 107–8, 109,
114, 125, 147, 148, 149,
151, 214
early days, 28
on legal cars, 132
on retiring, 146
profile, 84–85
remembered by fan, 121
Waltrip, Michael, 18, 151
profile, 86–87
Wannamaker, Dave, 125
Watkins Glen International, 52,
54, 131, 133, 220
area description, 117–19
fans camping at, 118
Weatherly, Joe, 27
West, Jerry, 185
Western Auto, 151
Wheeler, H. A. "Humpy," 152,
153, 190, 208, 214
profile, 88–90
White, Rex, 85

Whitehead, Kimberly, 1998
Miss Motorsports, 201
Williams, Hank, 184
Wilson, Bill, Earnhardt fan,
126
Wilson, Jim, 77
Winston 500, 18, 75–76
1998, 113–16
Winston Cup Awards Banquet,
1998, 191–93
Winston Cup championship,
146, 180
1992, 70
1996, 136
1998, 192
Winston Cup Media Tour, 1999,
88
Winston Cup racetracks, 220
Winston Cup Series, 51, 52, 54,
55, 56, 58, 62, 72, 74, 77,
81, 86, 104, 125, 129, 133,
143, 144, 152–53, 156, 161,
165, 167,176, 186–87, 193,
194, 208, 218
1992, 87
1993, 9, 149
1995, 134, 156
1996, 134
1997, 74, 116–17, 133
1998 qualifiers, 206–7
adding additional dates, 130
car design, in, 115–16
future of, 89–90, 140–41
Leader Bonus program, 134
media attitude, 21, 108, 109
Petty wins, 60
point system, 149–50
previously Grand National,
142
previously Strictly Stock Division, 144
traveling fans, 120
Winston Cup simulators, 200

Winston Open, The, 214
Winston Select 1995–1996,
    18
Winston West Series, 152
Winston, The
    1998, 213–15
    multiple events of, 214
Wolfe, Tom, *Esquire,* writer for,
    41
Wood Brothers, 86, 153, 160,
    214
Woody, Larry, 121
World Driving Champion,
    150

World of Outlaws races, 103
    1997, 194
WTBS, 209

Yarborough, Cale, 42, 44, 97,
    136, 182
Yastrzemski, Carl, 183, 185, 199
Yates, Robert, 11, 37, 39, 77, 81,
    144, 183
Yates, Rowdy, 183
Young, Henry, 124
Yunick, Smokey, 66

Zegar, Tara, 124

• • • • • • • • • • • • • • • the author

Monte Dutton is the motorsports writer at the *Gaston Gazette* in Gastonia, North Carolina. Dutton's work appears in "NASCAR This Week," a page that appears in more than 500 newspapers across the United States. He also writes a weekly column, "Monte Dutton on NASCAR." Both are syndicated by Universal Press Syndicate of Kansas City, Missouri. The author also works as Winston Cup correspondent for the weekly trade papers *FasTrack*, of Gastonia, North Carolina, and *Area Auto Racing News*, of Trenton, New Jersey, and for the GoCarolinas and Speednet web sites. His work has appeared in a number of magazines, including *Heartland USA, Racing Milestones* and *Inside NASCAR*. He is the winner of the Eastern Motorsports Press Association's Writer-of-the-year award for 1999. Monte Dutton is a graduate of Furman University (class of 1980) and lives in Clinton, South Carolina.